BARRON'S

Regents Exams and Answers

Earth and Space Sciences

Published by Kaplan North America, LLC, d/b/a Barron's Educational Series
1515 West Cypress Creek Road
Fort Lauderdale, Florida 33309
www.barronseduc.com

ISBN: 979-8-3497-0042-2

10 9 8 7 6 5 4 3 2

Contents

Chapter 1
How to Use This Book

Welcome to your complete guide for preparing for the 2025 Regents Examination in Earth and Space Sciences. This isn't just a test prep book—it's your all-in-one toolkit for mastering New York State's updated science standards and excelling on the new cluster-based Regents exam.

With real-world Earth and space sciences phenomena, Regents-style questions, and detailed explanations throughout, this book will help you learn how to think, analyze, and explain like a scientist—not just memorize facts. You'll engage with authentic scientific practices, interpret complex data, and build explanations grounded in evidence—just like the Regents exam expects you to.

What's Inside This Book

This book is organized into three main parts. Each is designed to help you build confidence step by step—from understanding the standards to applying your knowledge on full-length practice exams.

Chapters 2–5: Foundations of the Regents Earth and Space Sciences Exam

These chapters introduce everything you need to know about the 2025 Earth and Space Sciences Regents Exam.

Chapter 2: Understanding the Learning Standards and the 2025 Regents Earth and Space Sciences Exam

Learn what the New York State P–12 Science Learning Standards are and how they form the basis for everything on the Regents. You'll explore the three dimensions of science learning:

- Science and Engineering Practices (SEPs)—what scientists do
- Disciplinary Core Ideas (DCIs)—what scientists know
- Crosscutting Concepts (CCCs)—how science connects across domains

Chapter 3: Investigations—Learning Science by Doing Science

Discover how Investigations—hands-on, performance-based classroom tasks—help you develop scientific thinking skills that are also tested on the Regents exam. These tasks mirror real science, support instruction throughout the year, and strengthen your ability to analyze data, model systems, and explain natural Earth processes.

Chapter 4: The 2025 Regents Earth and Space Sciences Exam—Format, Topics, and What to Expect

Get familiar with how the test is structured: phenomenon-based clusters, question types, and topic breakdowns. You'll learn how many questions to expect, which Earth and space sciences topics are emphasized, and how to interpret graphs, maps, data tables, models, and crosscutting science concepts.

Chapter 5: Test-Taking Tips for the Regents Earth and Space Sciences Exam

Build a smart strategy for exam day. Learn how to tackle data-rich clusters, manage your time, and use the claim-evidence-reasoning (CER) method in written responses. You'll also get tips for decoding scientific models and avoiding common mistakes.

Chapter 6: Regents-Style Practice Clusters

This is the heart of the book—seven full practice clusters, one for each major topic on the test. Each cluster is built to match the format of the 2025 Regents Exam and includes:

- A real-world scientific phenomenon related to Earth or space science
- One or more data-rich stimuli (e.g., graphs, diagrams, satellite imagery, tables, or passages)
- 9-11 multiple-choice and constructed-response questions
- SEP, DCI, and CCC labels at the top of each cluster
- Detailed answer explanations for every question

Topics include:

- Earth's Place in the Universe
- Earth's Systems and Cycles
- Weather and Climate
- Human Impacts on Earth Systems
- History of Earth
- Natural Resources and Sustainability
- Engineering, Technology, and Applications of Science

Chapter 7: Regents Practice Exams

Finish your preparation with two complete Regents practice exams, fully aligned with the 2025 exam format.

Each includes:

- Stimulus-based question clusters modeled on the real exam
- A realistic mix of multiple-choice and short-response questions
- Detailed answer explanations for all items

Use these exams to:

- Benchmark your readiness
- Practice under timed conditions
- Target your final areas for review

How to Work Through This Book

Follow this step-by-step approach to get the most out of your study time.

Step 1: Learn the Foundations
Read Chapters 2–5 to understand how the Regents Earth and Space Sciences exam works, how to use science practices, and what's expected of you in a three-dimensional science assessment.

Step 2: Practice with Clusters
Choose a topic you're learning in class and work through the related cluster. Read the phenomenon carefully, study the stimuli (like diagrams, maps, or data tables), and answer each question thoughtfully using scientific reasoning.

Step 3: Write and Explain
Practice writing CER-style responses. Use Earth science vocabulary. Refer to models, geologic data, or satellite imagery when supporting your answers.

Step 4: Review Your Mistakes
Don't just check for right or wrong answers. Read the explanations to understand why each answer is correct and what misconceptions you should avoid in the future.

Step 5: Take the Full Practice Exams
When you're ready, simulate test-day conditions and take the practice exams in Chapter 7. Time yourself. Then go back and review your performance cluster by cluster.

Tips for Success

- Think like a geoscientist: ask questions, look for patterns in Earth systems, and explain how and why natural events occur.
- Use visual data: models, graphs, cross sections, and satellite images often hold key information for answering questions.
- Practice CER writing: make a clear claim, back it up with evidence from the data, and explain your reasoning using Earth science concepts.
- Review SEPs, DCIs, and CCCs to understand what science skills and big ideas each question is targeting.

Final Word

This book is designed to help you succeed—not by memorizing lists of terms but by helping you understand Earth and space sciences and apply it to the world around you. By the end, you'll be able to read like a scientist, think like a systems problem solver, and write like an expert. These are all skills that will help you not only on the Regents exam but in your future studies and life.

Chapter 2
Understanding the Learning Standards and the 2025 Regents Earth and Space Sciences Exam

Welcome to your first step in preparing for the Regents Examination in Earth and Space Sciences. This chapter is here to help you understand what you're expected to know, why you're learning it, and how it will show up on the exam. We'll walk through the New York State P-12 Science Learning Standards (NYSP-12SLS)—the foundation for your entire Earth and space sciences course—and show you how they connect to the test you'll take.

These standards guide everything from classroom investigations to Regents exam questions. They're built around the idea that science is about making sense of real-world phenomena, using data, models, and evidence to figure out how and why Earth systems work the way they do.

In this chapter, you'll learn how to decode the three dimensions of these standards, how they shape the structure of the 2025 exam, and what skills you'll need to think, communicate, and problem solve like a scientist.

Why Are There New Learning Standards?

The New York State P-12 Science Learning Standards were adopted to match national expectations and to help you develop deeper, more useful science skills. The new standards are based on the Next Generation Science Standards (NGSS), which aim to get students thinking, doing, and communicating like scientists—not just memorizing facts.

What Are the New Learning Standards?

The NYSP-12SLS are built on a three-dimensional learning model, which means every topic you study in Earth and Space Sciences will include:

1. Science and Engineering Practices (SEPs)
2. Disciplinary Core Ideas (DCIs)
3. Crosscutting Concepts (CCCs)

Let's break these down so you know what each means for you as a student.

Science and Engineering Practices (SEPs)

You're not just here to memorize facts—you're here to *think like a scientist*. That's what the Science and Engineering Practices (SEPs) are all about. They represent the key skills that real scientists and engineers use every day to investigate the world, solve problems, and build new technologies.

On the Regents Examination in Earth and Space Sciences, you'll be expected to use these skills—not just recognize them. That means interpreting data, analyzing geologic processes, modeling Earth systems, and explaining scientific ideas clearly. The exam is designed to mirror how science works in the real world.

Let's take a closer look at each SEP and how it shows up in your earth and space science studies and on the test.

1. Asking Questions and Defining Problems

What it means: Science starts with curiosity. Scientists ask questions about natural phenomena, while engineers define problems that need solutions.

In Earth science class: You might ask questions. Here are some examples: What causes earthquakes to occur in certain regions? How can coastal cities reduce flood risk?

On the test: You may be given a scenario and be asked to identify a testable question or propose a scientific investigation.

Example: "Based on observed changes in shoreline erosion, what question could a scientist ask to investigate the cause?"

2. Planning and Carrying Out Investigations

What it means: Scientists design experiments to test hypotheses. They choose variables, control conditions, collect data, and ensure reliability.

In Earth science class: You might conduct a lab experiment to see how different materials affect heat absorption or investigate soil permeability in various locations.

On the test: You could be asked to identify independent and dependent variables, describe controls, or analyze a proposed experimental design.

Example: "Design an investigation to test how surface material affects water runoff during rainfall."

3. Analyzing and Interpreting Data

What it means: Scientists use graphs, tables, and charts to make sense of what they observe. Analyzing data helps uncover patterns and draw conclusions.

In Earth science class: You might look at a graph showing global temperature trends or seismic activity over time.

On the test: You'll be asked to interpret visual data, find trends, and explain what the data show.

Example: "Use the graph to determine during which decade the rate of glacier melting increased most rapidly."

4. Constructing Explanations and Designing Solutions

What it means: Scientists explain why things happen based on evidence. Engineers use what they know to design solutions to real-world problems.

In Earth science class: You may explain how convection drives plate tectonics or how greenhouse gases contribute to climate change.

On the test: You might be asked to write a short explanation using scientific reasoning or propose a solution to an environmental problem.

Example: "Construct a scientific explanation for how deforestation contributes to increased atmospheric carbon dioxide."

5. Using Mathematics and Computational Thinking

What it means: Scientists use math to measure, calculate, and analyze data. Computational thinking helps with modeling systems or solving complex problems.

In Earth science class: You might calculate the rate of sea-level rise or use scaling to model planetary distances.

On the test: You could be asked to solve problems, interpret numerical data, or use ratios and percentages.

Example: "Using the data table, calculate the average rate of tectonic plate movement over the past 10 years."

6. Engaging in Argument from Evidence

What it means: Science involves discussion and debate. Scientists support their ideas with evidence and challenge ideas that aren't supported by data.

In Earth science class: You might argue for or against a claim using evidence from a climate report or rock sample data.

On the test: You'll be asked to evaluate claims, identify supporting evidence, or construct an argument of your own.

Example: "Based on the satellite data provided, argue whether the deforestation rate is contributing significantly to regional climate change."

7. Developing and Using Models

What it means: Models are tools scientists use to represent systems, explain processes, and make predictions. Models can be physical, visual, or conceptual.

In Earth science class: You may draw a model of Earth's interior layers or use a diagram to show the water cycle.

On the test: You'll be asked to interpret, use, or even modify a model to explain an Earth science concept.

Example: "Use the cross section of Earth's crust to identify where subduction is occurring and explain the resulting geological activity."

Why SEPs Matter

Every question on the 2025 Regents Earth and Space Sciences exam is designed to reflect real science. This means you won't just be recalling information—you'll be doing something with it. In fact, you will be doing the following:

- Analyzing investigations and field data
- Interpreting graphs, maps, and diagrams
- Applying knowledge to unfamiliar Earth systems and events
- Using models and constructing evidence-based explanations
- Justifying your answers with scientific data

How to Practice SEPs

To build your skills:

- Ask your own questions during lessons and labs
- Think critically when you read science articles or examine Earth data
- Explain your reasoning out loud or in writing
- Use visuals and models to study and summarize big ideas
- Get hands-on with labs, maps, or simulations whenever possible

Final Thought: Practice Makes a Scientist

The more you practice using these skills, the more confident you'll be—not just for the Regents exam but for understanding and exploring the dynamic Earth. Science is something you do. With these practices, you're well on your way.

Disciplinary Core Ideas (DCIs)

The Disciplinary Core Ideas (DCIs) are the heart and soul of what you'll study in Earth and space sciences. These are the core content areas—the essential topics that scientists agree every Earth science student should understand by the end of high school.

When you take the Regents Examination in Earth and Space Sciences, these are the major ideas you'll be tested on. The questions won't just ask for facts. They'll ask you to apply these ideas, use data, analyze models, and explain phenomena.

Each DCI focuses on different components of Earth systems and their interactions, from the motion of planets to human impacts on Earth. Let's break them down.

ESS1: Earth's Place in the Universe

What it's about: This core idea focuses on Earth's position in space and how celestial events (like seasons, eclipses, and moon phases) are influenced by the motion of Earth and other objects in the solar system.

What you'll need to know:

- The structure and scale of the solar system and universe
- Earth's rotation, revolution, and tilt
- Seasons, tides, eclipses, and moon phases
- Star formation and life cycles
- Fossil evidence and radiometric dating of Earth's history

Example: "Explain how Earth's tilt and orbit around the Sun cause seasonal changes in temperature and daylight hours."

ESS2: Earth's Systems

What it's about: Earth is a dynamic system with interacting components like the geosphere, atmosphere, hydrosphere, and biosphere. This DCI focuses on how these systems change over time and influence each other.

What you'll need to know:

- Plate tectonics, earthquakes, and volcanoes
- Weathering, erosion, and deposition
- The rock cycle and Earth's materials
- Ocean currents and atmospheric circulation
- The water cycle and climate systems

Example: "Describe how convection in the mantle contributes to plate movement and the formation of mountains."

ESS3: Earth and Human Activity

What it's about: Humans are part of Earth's systems, and our actions affect the planet's natural balance. This DCI explores how we use resources, impact the environment, and develop strategies for sustainability.

What you'll need to know:

- Natural resources (fossil fuels, minerals, water)
- Climate change and global warming
- Human impact on ecosystems and landscapes
- Natural hazards and disaster preparedness
- Solutions for environmental sustainability

Example: "Evaluate how the burning of fossil fuels contributes to global climate change and propose a possible mitigation strategy."

ESS1.C: The History of Planet Earth

What it's about: Earth has a long and complex history that scientists uncover using geologic evidence. This DCI includes fossils, layers of rock, and dating techniques used to reconstruct Earth's past.

What you'll need to know:

- Geologic time scale
- Stratigraphy and fossil records

- Radiometric dating
- Major extinction events and evolutionary changes
- Tectonic activity over Earth's history

Example: "Use the geologic column to determine the relative age of rock layers and explain how fossils support the idea of evolutionary change."

Earth and Life Science Connections

Why it matters: Earth's systems influence where and how life exists. Some Regents questions will connect Earth processes to living systems and biological change over time.

What you might see:

- Climate's role in species migration and survival
- Natural disasters affecting ecosystems
- Atmospheric changes influencing biodiversity

Example: "Explain how rising ocean temperatures might impact coral reef ecosystems and the species that depend on them."

Engineering and Math Connections

Why it matters: Scientists use engineering practices to solve Earth-related problems and math to model natural processes and analyze data. You'll be asked to calculate, graph, and design solutions.

What you might see:

- Interpreting satellite images or weather maps
- Modeling groundwater flow or erosion
- Using graphs to analyze sea level rise or carbon emissions

Example: "Design a solution to reduce coastal erosion and explain how your design addresses both environmental and human concerns."

How to Master the DCIs

To succeed on the Regents exam:

- Focus on *understanding,* not just memorizing.
- Practice applying concepts in new situations.
- Use models, data sets, and real-world examples.
- Study how the DCIs connect to each other (for example, how Earth's systems influence human activity).

Final Thought: The Big Picture

Each DCI represents a major branch of Earth and space sciences. Together, they help you make sense of the planet—how it works, how it changes, and how we can protect it. Whether you're studying how tectonic plates shape continents, how climate affects ecosystems, or how humans impact natural resources, the Disciplinary Core Ideas give you the tools to think scientifically and ace the Regents exam.

Crosscutting Concepts (CCCs)

Science isn't just about memorizing facts—it's about thinking like a scientist. That means being able to make connections across different topics and disciplines. That's where Crosscutting Concepts (CCCs) come in. These big-picture ideas are used by scientists in every field, and they help you see patterns and relationships in the world around you.

Each CCC is like a lens you can use to examine a scientific problem. On the Regents Earth and Space Sciences exam, you'll be expected to apply these concepts to real-world data, situations, and problems. Let's explore each one.

1. Patterns

What it means: Scientists look for repeating patterns to make predictions and discover relationships. In Earth science, patterns appear in climate trends, earthquake locations, rock formations, and planetary motion.

On the test: You might be shown a data set or map and be asked to identify a pattern, such as earthquake zones along plate boundaries or changes in temperature over time.

Example: "Based on the pattern of seismic activity shown on the map, where is an earthquake most likely to occur next?"

2. Cause and Effect

What it means: Every action has a consequence. Understanding cause and effect helps scientists figure out why natural events happen and what might result.

On the test: You may have to explain how one Earth system influences another—for example, how volcanic eruptions affect atmospheric temperatures.

Example: "Describe the cause-and-effect relationship between increased greenhouse gas emissions and rising global temperatures."

3. Scale, Proportion, and Quantity

What it means: Earth processes occur at many scales, from local weather patterns to global climate systems. Scientists use measurements and calculations to compare and understand these processes.

On the test: You might analyze maps or data to compare sizes, time scales, or rates of change.

Example: "Compare the time scales of weathering and erosion in forming different types of landforms."

4. Systems and System Models

What it means: Scientists often model complex systems—like the atmosphere, hydrosphere, and geosphere—to study how different parts interact.

On the test: You could be asked to interpret or modify a model of an Earth system, such as the water cycle or plate tectonics.

Example: "Use a model to show how the movement of tectonic plates creates earthquakes and mountain ranges."

5. Energy and Matter

What it means: Energy drives many Earth processes, and matter cycles through the planet's systems. Understanding how both move helps explain everything from the rock cycle to climate change.

On the test: You'll analyze energy transfer (e.g., solar energy, convection currents) or matter cycling (e.g., carbon or water cycles).

Example: "Explain how energy from the Sun powers wind and ocean current systems."

6. Structure and Function

What it means: The shape or structure of something often tells you how it works. This concept applies to things like landforms, mineral crystals, and atmospheric layers.

On the test: You'll connect the structure of natural features or systems to their roles in Earth's processes.

Example: "Describe how the layered structure of the atmosphere affects the movement of air and weather patterns."

7. Stability and Change

What it means: Some Earth systems stay stable over time, while others change suddenly or gradually. Scientists study both types to understand long-term and short-term processes.

On the test: You may be asked to explain how Earth systems remain balanced or how they shift due to events like climate change or earthquakes.

Example: "Explain how feedback mechanisms in the climate system can either stabilize or amplify global temperature changes."

Putting It All Together

Often, more than one crosscutting concept will apply to a single question. For example, you might examine patterns in data while also describing a cause-and-effect relationship and considering energy flow in a system.

Tip: When practicing, ask yourself:

- What pattern do I see?
- What is causing this to happen?
- What is the role of structure or energy here?
- How does this system work as a whole?

Being able to think in these ways—not just memorize content—is key to acing the new Regents exam.

Chapter 3

Investigations—Learning Science by Doing Science

As a student preparing for the Regents Examination in Earth and Space Sciences, you're probably used to studying facts, reading explanations, and answering practice questions. But did you know that doing science—through hands-on investigations—is just as important as reading about it?

This chapter introduces you to a new and exciting part of science learning in New York State: Investigations. These aren't quizzes or multiple-choice tests. They're performance-based science tasks that give you the chance to think, work, and reason like a real Earth scientist or engineer.

Let's explore what Investigations are, how they help you build important skills, and what they look like in your classroom.

What Are Investigations?

Real Science, Not Just a Lab Report

Investigations are hands-on, minds-on science activities that are now part of the high school Earth and space sciences experience across New York State. These tasks are meant to be authentic, meaning they resemble the way scientists and engineers actually solve problems in the real world. Investigations can be described as follows:

- **Curriculum embedded.** They are done in class as part of your learning.
- **Locally administered.** They are created and scored by your teachers or district.
- **Performance based.** You show what you know by doing, not just by writing.
- **Aligned to the state's science standards.** They help you prepare for the Regents exam.

Note: Investigations are *not part of the Regents exam itself*, but they help you develop the same kinds of *scientific thinking and reasoning skills* that the Regents exam will assess.

Why Are Investigations Important?

Think of Investigations as practice for real-world science. You get to explore geological and earth science phenomena, ask questions, test ideas, analyze data, and communicate your findings— just like a professional geologist, meteorologist, oceanographer, or climate scientist.

How Investigations Help You

Investigations reinforce the three dimensions of science learning:

- Science and Engineering Practices (SEPs)—what scientists do
- Disciplinary Core Ideas (DCIs)—what scientists know
- Crosscutting Concepts (CCCs)—how ideas connect across science

Investigations develop your ability to:

- Plan and carry out experiments
- Build and use models
- Interpret graphs and tables
- Support your explanations with data
- Think critically and reason through scientific problems

These are the exact skills you'll need for the Regents clusters, especially the constructed-response questions where you'll be asked to make claims, analyze patterns, or evaluate models.

What Does an Investigation Look Like?

Let's walk through a few example scenarios so you know what to expect.

Example 1: Modeling Plate Tectonics

Scenario: You're given a cross-section model showing converging tectonic plates and asked to predict what geological features will form.

What you might do:

- Use diagrams to identify subduction zones and plate boundaries
- Explain the formation of trenches, mountains, or volcanoes
- Compare your predictions to real-world tectonic activity data

Skills used: Developing and using models | Constructing explanations | Analyzing data

How it helps: This mirrors the kinds of data interpretation questions you'll see on clusters like "Earth's Systems."

Example 2: Investigating Ocean Currents and Temperature

Scenario: You examine how different water temperatures affect ocean current flow in a simulation or lab setup.

What you might do:

- Design a fair test by controlling salinity and water depth
- Measure current strength or direction in warm vs. cold water
- Graph your results and explain the role of convection in driving ocean currents

Skills used: Planning and carrying out investigations | Using math and computational thinking

How it helps: This reinforces concepts from "Weather and Climate" and builds experience interpreting physical data—just like on the Regents.

Example 3: Ecosystem Change—Deforestation and Carbon Cycling

Scenario: You're given satellite imagery and data showing carbon levels in a region before and after deforestation.

What you might do:

- Analyze carbon levels in the atmosphere and soil
- Explain how the removal of vegetation impacts the carbon cycle
- Construct a model of the system before and after deforestation

Skills used: Interpreting models | Arguing from evidence | Cause and effect reasoning

How it helps: This connects directly to the type of reasoning used in "Earth and Human Activity" clusters on the Regents.

Investigations vs. Regents Clusters

Investigations	Regents Clusters
Hands-on tasks done in class	Written tasks done during the state exam
Locally scored and flexible in timing	Standardized and scored by the state
Focus on applying science and problem solving	Focus on analyzing data and explaining reasoning
Real-time exploration and discussion	Independent written responses under timed conditions
Build the same skills tested on the Regents	Assess those skills through questions and CER tasks

Final Thoughts: Why This Matters to You

Investigations are more than just school assignments—they're your training ground for thinking scientifically. By participating in classroom investigations, you'll gain confidence in:

- Analyzing data
- Making sense of real-world problems
- Explaining Earth and space sciences phenomena clearly and with evidence

These experiences will make the Regents exam feel familiar, not intimidating. You'll be better prepared to tackle data-rich clusters, models, and constructed-response questions because you've already practiced those skills in a meaningful way. So when your teacher says, "Today we're doing an Investigation," take it seriously—and give it your best. You're not just preparing for a class grade. You're training to succeed on the Regents.

Chapter 4

The 2025 Regents Earth and Space Sciences Exam—Format, Topics, and What to Expect

You've learned about the three dimensions of the New York State P–12 Science Learning Standards. Now it's time to get specific. What's actually on the Regents Examination in Earth and Space Sciences? How is the test set up? What topics are covered? And how much should you focus on each?

This chapter answers those questions and more—so you know exactly what to expect on exam day.

Test Format Overview

The new Regents Earth and Space Sciences Exam is three hours long and will include both multiple-choice and constructed-response questions. Here's how it breaks down:

Section	Question Type	Estimated % of Exam
Multiple-Choice Questions	Choose from 4 answer choices	~60%
Constructed-Response Questions	Write your own answers (short answers, data analysis, models, etc.)	~40%

You'll answer 45–55 questions, grouped into 9 to 11 clusters (groups of related questions).

How Questions Are Organized: Clusters and Storylines

Unlike older exams, the new Regents Earth and Space Sciences exam is built around question clusters. Each cluster:

- Is based on a real-world scientific phenomenon
- Includes multiple stimuli (e.g., graphs, maps, cross sections, data sets, satellite images, passages)

- Asks related questions that build toward an explanation, a model, or a solution
- Includes both multiple-choice and constructed-response items

Think of each cluster as a mini science story. You'll analyze information, interpret data, and explain your thinking just like a real Earth scientist.

Topic Breakdown: How Much of Each Topic Is on the Test?

The chart below shows the topic-level test blueprint from the NYSED Educator Guide. It tells you what percentage of the exam is focused on each topic area:

Topic	% of Questions
Earth's Place in the Universe	6–16%
Earth's Systems	27–41%
Weather and Climate	13–23%
Earth and Human Activity	13–23%
History of Earth	6–16%
Earth's Systems (Crosscutting integration focus)	5–11%
Engineering, Technology, and Applications of Science	3–11%

What Should You Expect on the Exam?

On the exam, you'll see different types of stimuli and have to answer different types of questions. You will be allowed to use a calculator when taking the exam—but only specific types of calculators.

Types of Stimuli You'll See

Every question cluster will include stimuli, or real-world data and visuals, such as:

- Graphs and data tables
- Diagrams and scientific models
- Maps, cross sections, and satellite images
- Photos and real case studies
- Short reading passages

You'll use these materials to answer related questions that assess your ability to *interpret information, analyze patterns,* and *construct explanations.*

Question Types You'll Answer

You'll face two kinds of questions throughout the exam: multiple-choice questions and constructed-response questions.

Multiple-Choice Questions

- Standard format: choose the best answer from four options
- Tests recall, interpretation, and basic application skills

Constructed-Response Questions

- Short answers that require writing
- May ask you to interpret data, draw conclusions, support claims with evidence, or develop models
- Designed to show how well you understand and apply what you've learned in Earth and space sciences

Calculator Use

You may use a four-function or scientific calculator during the exam. Graphing calculators are NOT allowed.

Final Takeaways

- The 2025 Regents Earth and Space Sciences exam is built around real-world phenomena and scientific reasoning, not just memorization.
- Expect a balanced mix of multiple-choice questions and constructed-response questions, with questions organized into story-based clusters.
- Practice using stimuli—like data tables, graphs, maps, and models—to answer questions just as you'll do on the real test.

Chapter 5

Test-Taking Tips for the Regents Earth and Space Sciences Exam

The Regents Earth and Space Sciences exam is not just about memorizing facts. It's about applying your knowledge to real-world science problems, interpreting data and visuals, and explaining your reasoning. This chapter provides tips to help you feel calm, focused, and ready to tackle the exam with confidence.

Before the Test: Set Yourself Up for Success

1. Understand the Exam Format

- You'll answer questions in clusters (mini science scenarios).
- Each cluster includes stimuli (charts, maps, models, or passages) and 9–11 related questions.
- The exam includes:
  - Multiple-choice questions (about 60%)
  - Constructed-response questions (about 40%)
- Total test time: 3 hours

2. Know the Three Dimensions

- Every question is based on:
  - A Science and Engineering Practice (SEP)—what scientists do
  - A Disciplinary Core Idea (DCI)—what scientists know
  - A Crosscutting Concept (CCC)—how science connects ideas
- Get familiar with what each one looks like in a question.

3. Study Strategically

- Review each topic cluster in this book.
- Practice explaining phenomena and models using scientific vocabulary.
- Create flashcards for key vocabulary and processes (e.g., convection currents, greenhouse effect).

- Practice reading graphs, tables, and models. This is a skill you'll use on almost every question cluster.

During the Test: Smart Strategies for Every Question

1. Read the Stimulus Carefully

- Take your time reading the phenomenon and accompanying data before jumping to the questions.
- Underline or circle important information (e.g., trends in a graph, key terms in a map, cause-effect words in a paragraph).

2. Don't Rush the Clusters

- Questions are related and often build in complexity.
- Use information from earlier questions to help with later ones.
  Example: If a graph shows rising global temperatures, a later question might ask how polar ice melt or ocean currents are affected.

3. Eliminate Wrong Answers in Multiple-Choice Questions

- Cross out clearly wrong answers.
- Look for key words like *always, only,* or *most likely*. These can help identify traps.
- If stuck, make an educated guess. There's no penalty for guessing.

4. Master Constructed Responses

- Use the CER (claim-evidence-reasoning) method when appropriate:
  - Claim—a direct answer to the question
  - Evidence—data or information from the stimulus
  - Reasoning—the science behind why the evidence supports the claim
- Use scientific vocabulary (e.g., subduction, convection, carbon cycle).
- Write in complete sentences. Avoid writing "I think" or "I believe." Be confident in your science!

5. Manage Your Time

- Don't get stuck too long on any one question.
- Aim to spend no more than 15–20 minutes per cluster.
- If a question seems difficult, skip it and come back later.

Bonus Tips for High-Scoring Answers

- Use models and data in your explanation. Refer directly to graphs, maps, or tables ("As shown in Figure 2 . . .").
- Label diagrams if asked to draw one. Use arrows, keywords, and clear titles.
- Review your writing. Check that your answers match what the question asked—especially for multipart questions.

Calm Your Nerves: Test Day Checklist

- Get a good night's sleep.
- Eat a healthy breakfast.
- Bring a pen, pencil, and a permitted calculator.
- Breathe deeply and stay focused.
- Use positive self-talk: "I've practiced this. I know how to read models. I've got this!"

Final Advice

Remember: this test isn't about memorizing every science fact. It's about thinking like a scientist. That means:

- Reading carefully
- Explaining your thinking
- Applying knowledge to new situations
- Looking for patterns and evidence

The Regents Earth and Space Sciences exam is your chance to show how much you've grown as a science learner. With the right strategies, mindset, and preparation, you can excel.

Good luck. You're ready!

Chapter 6

Regents-Style Practice Clusters

What's This Chapter About?

This chapter features original question clusters specifically designed to mirror the format of the Regents Examination in Earth and Space Sciences. These clusters are based on the New York State P-12 Science Learning Standards and use a three-dimensional approach that combines:

- Science and Engineering Practices (SEPs)—what scientists do
- Disciplinary Core Ideas (DCIs)—the big ideas in Earth and space sciences
- Crosscutting Concepts (CCCs)—themes that connect all sciences

Rather than testing isolated facts, the Regents exam asks you to analyze real-world scientific situations using models, data, graphs, maps, and case studies.

How to Use This Chapter

You'll find one complete question cluster for each of the 7 major topic areas tested on the exam:

1. Earth's Place in the Universe
2. Earth's Systems and Cycles
3. Weather and Climate
4. Human Impacts on Earth Systems
5. History of Earth
6. Natural Resources and Sustainability
7. Engineering, Technology, and Applications of Science

Each cluster includes:

- A real-world phenomenon
- 9–11 questions (just like the actual exam)
- Detailed answer explanations at the end of each cluster

Your Goal

Use these clusters to sharpen your skills in thinking, reasoning, and explaining like an Earth scientist. As you work through each one:

- Read the phenomenon carefully.
- Analyze the diagrams, maps, and data.
- Use evidence to explain your reasoning.
- Check your answers and learn from your mistakes.

Topic 1: Earth's Place in the Universe

Science and Engineering Practice (SEP): Developing and Using Models

Disciplinary Core Idea (DCI): LS3.A: ESS1.B: Earth and the Solar System

Crosscutting Concept (CCC): Patterns

Phenomenon: Changing Visibility of Constellations Throughout the Year

Base your answers to questions 1 through 9 on the information below and on your knowledge of Earth and space sciences.

Seasonal Visibility of Constellations

People have observed patterns in the night sky for thousands of years. One such pattern is that some constellations are visible only during certain seasons. For example, Orion is prominent in winter, while Scorpius is visible in summer. This phenomenon occurs due to Earth's orbit around the Sun.

Sample Constellations and Visibility by Season

Constellation	Primary Visibility
Orion	Winter
Scorpius	Summer
Leo	Spring
Pegasus	Fall

1. Which conclusion is best supported by the data in the table?
   (1) The same constellations are visible all year long.
   (2) Earth's rotation causes stars to appear in different seasons.
   (3) Different constellations are visible in different seasons due to Earth's orbit.
   (4) The Moon's phases control which stars are visible.

2. Using the data, explain why certain constellations are visible only during specific seasons. Include Earth's movement in your response.

____

____

____

____

Model of Earth's Orbit and Night Sky Visibility

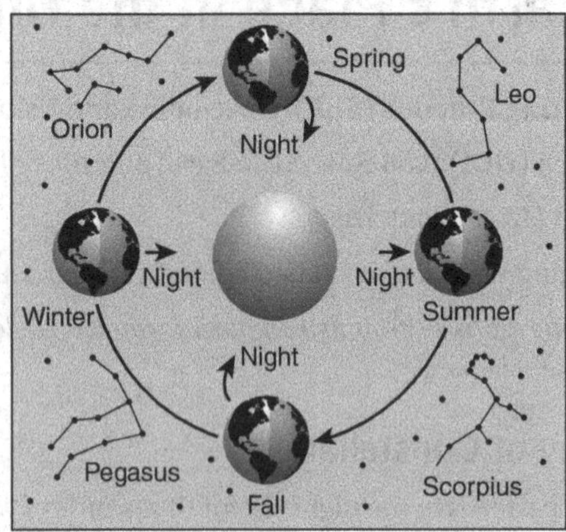

3. Which motion is primarily responsible for the change in visible constellations over the year?
   (1) Rotation of Earth on its axis
   (2) Revolution of Earth around the Sun
   (3) Moon orbiting Earth
   (4) Rotation of the Sun

4. Based on the model, explain how Earth's position in its orbit affects which constellations are visible at night.

   ____

   ____

   ____

   ____

Daily vs. Yearly Patterns in the Sky

While Earth's rotation causes stars to rise and set each night, Earth's revolution causes long-term shifts in star visibility over the course of the year.

5. Which event is caused by Earth's rotation, not its revolution?

(1) Orion is visible in winter but not in summer.
(2) The Sun rises in the east and sets in the west each day.
(3) Different constellations appear at different seasons.
(4) The Moon's phases repeat in a 29-day cycle.

6. Describe how Earth's daily rotation and yearly revolution work together to create observable patterns in the night sky.

____

____

____

____

Ancient Uses of Sky Patterns

Ancient cultures used the appearance of seasonal constellations to guide agriculture, navigation, and calendars. Some civilizations built monuments aligned with celestial events like solstices or the appearance of key constellations.

7. Which of the following best explains how patterns in the sky were used historically?

(1) To measure ocean depth
(2) To forecast earthquakes
(3) To track time and seasonal changes
(4) To predict solar wind activity

8. Give one example of how ancient civilizations may have used patterns in star visibility to benefit their society.

____

____

____

____

9. Claim: "Earth's motion through space creates predictable patterns in the sky."

Using examples from the cluster, construct a well-supported argument to explain how Earth's revolution and rotation produce these patterns.

___

___

___

___

Answer Explanations

1. **(3)** The data in the table show that specific constellations are visible only during particular seasons—for example, Orion in winter and Scorpius in summer. This change in visibility is not due to Earth's daily rotation (which causes stars to rise and set over a single night). Rather, this change in visibility is due to Earth's yearly revolution around the Sun, which gradually shifts the direction Earth faces at night. As a result, different constellations are visible in the night sky at different points in Earth's orbit.
2. **Sample Response:** The data in the table show that specific constellations are visible only during particular seasons—for example, Orion in winter and Scorpius in summer. This change in visibility is not due to Earth's daily rotation (which causes stars to rise and set over a single night). Rather, this change in visibility is due to Earth's yearly revolution around the Sun, which gradually shifts the direction Earth faces at night. As a result, different constellations are visible in the night sky at different points in Earth's orbit.
3. **(2)** The changing visibility of constellations throughout the year is a result of Earth's revolution around the Sun. This orbital motion causes our planet to face different parts of space at night during each season. In contrast, Earth's rotation (spinning on its axis) causes stars to rise in the east and set in the west each day, but it doesn't change which constellations are visible across months.
4. **Sample Response:** In the model, Earth's orbit is divided into four positions—one for each season. At each position, the "night side" of Earth faces away from the Sun and toward a different part of space. For example, in winter, Earth's night side faces the direction of Orion, while in summer, it faces Scorpius. These differences in viewing direction are a direct result of Earth's orbital motion. Because Earth orbits the Sun in a predictable pattern, the constellations we see from the night side also follow predictable seasonal patterns.
5. **(2)** The daily rising and setting of celestial objects like the Sun and stars is caused by Earth's rotation on its axis, which takes approximately 24 hours. This spinning motion causes the Sun to appear to move across the sky from east to west and is responsible for the cycle of day and night. In contrast, the revolution of Earth governs seasonal changes.
6. **Sample Response:** The daily rising and setting of celestial objects like the Sun and stars is caused by Earth's rotation on its axis, which takes approximately 24 hours. This spinning motion causes the Sun to appear to move across the sky from east to west and is responsible for the cycle of day and night. In contrast, the revolution of Earth governs seasonal changes.

7. **(3)** Throughout history, many civilizations observed patterns in the sky to track time and seasonal changes. For example, the rising of certain constellations marked the beginning of planting or harvesting seasons. People also used sky patterns to create early calendars and align religious festivals with celestial events. These uses were based on repeated, predictable sky patterns caused by Earth's motion.

8. **Sample Response:** Ancient Egyptians used the annual appearance of the star Sirius just before sunrise to predict the flooding of the Nile River, which was crucial for agriculture. The Mayans tracked the movement of Venus to plan ceremonial events. In many cultures, the rising of specific constellations helped farmers know when to plant or harvest crops. These examples show how knowledge of sky patterns was essential for survival and planning in early societies.

9. **Sample Response:** Earth's rotation and revolution create predictable patterns in the sky. The rotation of Earth causes the Sun and stars to rise in the east and set in the west each day. The revolution of Earth around the Sun changes the part of the sky we see at night over the course of a year. For example, Orion is visible in winter, while Scorpius appears in summer. These patterns are consistent and reliable, which is why ancient cultures used them for navigation, farming, and timekeeping. Together, Earth's motions shape the rhythms of the sky that we observe today.

Topic 2: Earth's Systems and Cycles

Science and Engineering Practice (SEP): Constructing Explanations and Designing Solutions

Disciplinary Core Idea (DCI): ESS2.A: Earth's Materials and System

Crosscutting Concept (CCC): Structure and Function

Phenomenon: Earthquake Activity Along Plate Boundaries

Base your answers to questions 1 through 9 on the information below and on your knowledge of Earth and space sciences.

Investigating Earthquakes and Plate Movement

Earthquakes occur when stress builds up and is suddenly released along faults in Earth's crust. These faults are most often found at the boundaries between tectonic plates. Scientists use seismographs and GPS data to monitor and predict earthquake risk in different regions.

A team of geologists studied seismic activity along a section of the San Andreas Fault. They recorded the number and magnitude of earthquakes over time and correlated this with observed plate movement.

Earthquake Activity Along the San Andreas Fault

Year	Number of Earthquakes (Magnitude 4.0+)	Average Plate Movement (cm/year)
1990	18	3.2
1995	25	3.5
2000	32	4.0
2005	29	4.3
2010	37	4.8

1. What conclusion is best supported by the data in the table?

(1) Earthquake frequency decreases as plate movement increases.

(2) Earthquakes occur more often when plate movement speeds up.

(3) Plate boundaries are not related to earthquake activity.

(4) Earthquakes reduce tectonic plate motion.

2. Using evidence from the table, explain the relationship between plate movement and earthquake frequency. Include how stress builds and releases along faults.

____

____

____

____

Structure of Earth's Crust and Plate Boundaries

The lithosphere is broken into tectonic plates that float on the semifluid asthenosphere. Where plates interact, stress builds up in rocks until it's released as seismic waves.

Cross Section of a Transform Plate Boundary (San Andreas Fault)

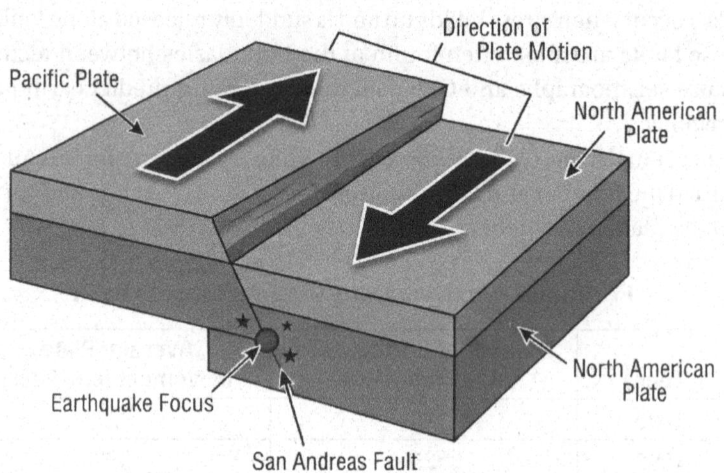

3. What most likely explains the frequent earthquake activity at transform boundaries like the San Andreas Fault?

(1) Subduction of one plate under another causes lava to erupt.

(2) Compression forces push land upward to form mountains.

(3) Stored stress from plate friction is suddenly released.

(4) Plates move apart, creating a rift valley.

4. Using the model above, describe how plate movement leads to earthquakes. Explain how the structure and motion of plates affect the release of energy.

____

____

____

____

Measuring Earthquake Impacts

Scientists use the moment magnitude scale to measure energy released during an earthquake. Ground motion data are collected by seismometers, which help engineers assess structural damage risk and guide building design.

Estimated Ground Motion for Earthquakes of Varying Magnitudes

Magnitude	Ground Movement (mm)
4.0	0.1
5.0	1.0
6.0	10.0
7.0	100.0

5. Which statement best describes the relationship shown in the table?

(1) Ground motion increases tenfold with each whole-number increase in magnitude.

(2) Ground motion decreases as magnitude increases.

(3) There is no consistent pattern between magnitude and ground motion.

(4) Earthquakes of all magnitudes release the same energy.

6. Use the table to explain how even small changes in earthquake magnitude can significantly increase ground motion and damage potential.

____

____

____

____

Engineering for Earthquake Safety

A structural engineer designs a bridge in a high-risk earthquake zone. The engineer uses flexible joints and base isolators to reduce damage from seismic waves.

7. Which feature of the bridge design best illustrates how structure can be adapted to function in an earthquake-prone area?

(1) Rigid, anchored supports to hold the structure still
(2) Sharp angles and tall columns to increase stiffness
(3) Flexible components that absorb ground motion
(4) Cement flooring to prevent movement

8. Explain how understanding ground motion and plate behavior helps engineers design safer buildings in earthquake zones.

____

____

____

____

9. Claim: "The structure of Earth's crust and the way plates move determine where and how earthquakes happen."

Using examples from the cluster, construct a well-supported argument to explain how structure affects function in Earth systems.

____

____

____

____

Answer Explanations

1. **(2)** The data show that as average plate movement increases over time, the number of earthquakes also increases. This suggests a positive correlation between faster plate movement and earthquake frequency.
2. **Sample Response:** As plate movement increases, more stress builds along the fault lines. When the stress becomes too great, it's released in the form of an earthquake. The faster the plates move, the more frequently stress accumulates and triggers seismic events.
3. **(3)** At transform boundaries, plates slide past each other, causing stress to build. When friction is overcome, the stress is released suddenly, producing earthquakes.
4. **Sample Response:** In a transform boundary, tectonic plates move in opposite directions. This motion creates stress at the fault line, shown in the model. When the rocks can no longer withstand the pressure, they fracture and shift suddenly, releasing energy as seismic waves.
5. **(1)** Each whole-number increase on the magnitude scale corresponds to about a tenfold increase in ground movement. This exponential relationship explains why a 7.0 quake is far more destructive than a 5.0 quake.
6. **Sample Response:** A magnitude 5.0 earthquake causes 10 times more ground motion than a 4.0. By the time an earthquake reaches magnitude 7.0, ground motion has increased 1,000 times. This means the damage risk grows rapidly even with small increases in magnitude.
7. **(3)** Flexible joints and base isolators are designed to absorb and dissipate energy, allowing the structure to move safely with the ground during an earthquake rather than resist it.
8. **Sample Response:** Engineers use ground motion data to design buildings that can withstand shaking. By modeling how energy moves through Earth's crust, they create features like shock absorbers or reinforced frames to minimize damage and protect lives.
9. **Sample Response:** Earthquakes occur along faults where tectonic plates interact. The structure of Earth's crust—broken into moving plates—determines where stress accumulates. The way plates move (slide, collide, separate) influences the type and frequency of seismic activity. This shows how the physical structure of Earth systems controls their function and behavior.

Topic 3: Weather and Climate

Science and Engineering Practice (SEP): Analyzing and Interpreting Data

Disciplinary Core Idea (DCI): ESS2.D: Weather and Climate

Crosscutting Concept (CCC): Cause and Effect

Phenomenon: Global Ocean Currents and Coastal Climate Differences

Base your answers to questions 1 through 9 on the information below and on your knowledge of Earth and space sciences.

The Gulf Stream and Climate in Europe

The Gulf Stream is a warm ocean current that flows from the Gulf of Mexico across the Atlantic Ocean toward Europe. Scientists have long noted that cities in Europe at the same latitude as cities in North America tend to have milder winters.

Average Winter Temperatures at Similar Latitudes

City	Latitude	Average January Temperature (°C)
London, U.K.	51.5° N	5.0
Calgary, Canada	51.0° N	−7.0
Paris, France	48.8° N	4.0
Minneapolis, U.S.	45.0° N	−9.0

1. Which conclusion is best supported by the data in the table?
   (1) All cities at the same latitude have identical climates.
   (2) Temperature decreases only with increasing latitude.
   (3) Europe's coastal cities are warmer than inland cities at similar latitudes.
   (4) Ocean currents do not influence regional temperatures.

2. Using the data in the table, explain how the Gulf Stream affects the climate of cities in western Europe compared with those in North America.

____

____

____

____

How Ocean Currents Work

Ocean currents move energy around the planet. Warm water from the equator flows toward the poles, while cold water sinks and returns toward the equator. These currents are influenced by wind, Earth's rotation (Coriolis effect), and the shape of continents.

Simplified Diagram of the Gulf Stream

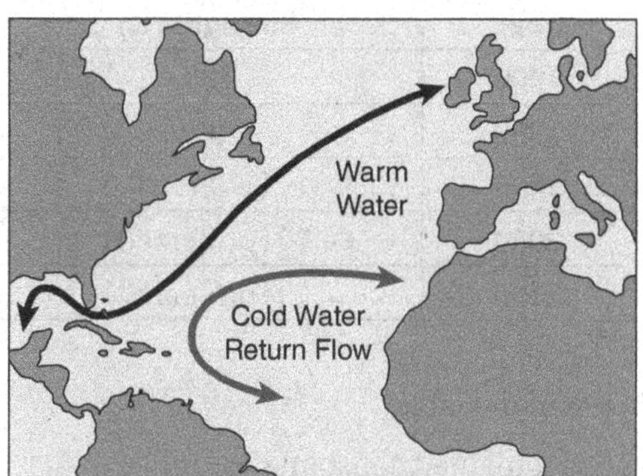

The Gulf Stream carriers warm water from the tropics to Europe's western coast, transferring heat to the atmosphere and affecting climate.

3. Which process is primarily responsible for driving surface ocean currents like the Gulf Stream?

(1) Plate tectonics

(2) Solar radiation and wind patterns

(3) Magnetic field shifts

(4) Precipitation and evaporation

4. Using the diagram above, describe how the structure of ocean circulation helps regulate Earth's climate.

____

____

____

____

Climate Change and Ocean Circulation

Recent data show that the Atlantic Meridional Overturning Circulation (AMOC), which includes the Gulf Stream, has weakened over the past century. Climate models suggest that continued warming and melting of polar ice could disrupt this system.

Simulated AMOC Strength Over Time

Year	Relative Circulation Strength (%)
1900	100
1950	95
2000	85
2050	70 (projected)
2100	60 (projected)

5. What trend is shown in the table?

(1) The AMOC has remained constant over time.

(2) The AMOC is gradually strengthening due to warming.

(3) The AMOC is weakening and is projected to continue declining.

(4) The AMOC changes only every 100 years.

6. Use the table to explain how changes in the AMOC could affect global weather and climate systems.

____

____

____

____

Feedback and Climate Systems

Ocean currents are part of a complex feedback system involving ice melt, sea level rise, and atmospheric patterns. When polar ice melts, it adds freshwater to the ocean, potentially disrupting salinity-driven circulation.

7. Which of the following best describes a feedback loop in the climate system?

(1) An isolated event with no long-term effects

(2) A one-way process that reverses itself naturally

(3) A chain reaction where change in one part influences other parts

(4) A process that occurs only during storms

8. Explain how melting polar ice could create a feedback loop that weakens ocean circulation and affects climate in coastal regions.

____

____

____

____

9. Claim: "Ocean currents are essential for distributing heat and shaping global climate patterns."

Using examples from the cluster, construct a well-supported argument to explain how ocean circulation contributes to weather and climate.

____

____

____

____

Answer Explanations

1. **(3)** Europe's coastal cities (like London and Paris) are significantly warmer than North American cities at similar latitudes (like Calgary and Minneapolis). This pattern supports the idea that proximity to the warm Gulf Stream current moderates temperatures in western Europe.
2. **Sample Response:** The data show that London (51.5° N) and Paris (48.8° N) have average winter temperatures well above freezing, while Calgary and Minneapolis—located at similar latitudes—experience much colder winters. This is because the Gulf Stream brings warm water from the tropics to Europe, which warms the air above the ocean. The warm air influences coastal climates, making them milder than inland regions without ocean current influence.
3. **(2)** Surface ocean currents are primarily driven by wind patterns and the distribution of solar energy, which heats water near the equator and sets air in motion. The Coriolis effect and continental positioning also influence the direction and flow of these currents.
4. **Sample Response:** The diagram shows warm water moving from the equator toward Europe and cold water sinking and returning to the tropics. This circulation helps distribute heat across the globe. As warm water transfers energy to the air, coastal regions like western Europe experience milder climates. Without this movement, heat would stay trapped in the tropics and polar areas would be much colder.
5. **(3)** The data in the table clearly show a decline in circulation strength over time. From 100% in 1900 to a projected 60% by 2100, the AMOC (which includes the Gulf Stream) is weakening, likely due to climate change and melting polar ice.
6. **Sample Response:** A weaker AMOC means less warm water is transported to the North Atlantic. This could lead to cooler temperatures in western Europe, stronger storms, and changes in tropical rainfall. It may also disrupt monsoon patterns and reduce the ocean's ability to absorb atmospheric carbon dioxide, intensifying climate change.
7. **(3)** A feedback loop involves a system where a change in one part affects other parts, which in turn loops back to influence the original system. For example, melting ice lowers salinity, weakening currents that would typically help regulate ice formation—creating a reinforcing cycle.

8. **Sample Response:** When polar ice melts, it adds freshwater to the ocean, reducing salinity and making the water less dense. This disrupts the sinking of cold water that helps drive ocean circulation. As a result, warm water may not be transported as efficiently, which can cool regions like Europe and alter global precipitation patterns—creating a feedback loop that accelerates climate change.
9. **Sample Response:** Ocean currents like the Gulf Stream move heat from the tropics to the poles, shaping regional climates. For example, they help western Europe stay warmer than North America at the same latitude. Ocean currents also influence storm systems and precipitation by affecting air temperature and pressure. As shown in the cluster, changes in ocean circulation due to melting ice could disrupt these climate patterns, making ocean currents essential to Earth's energy balance.

Topic 4: Human Impacts on Earth Systems

Science and Engineering Practice (SEP): Constructing Explanations and Designing Solutions

Disciplinary Core Idea (DCI): ESS3.C: Human Impacts on Earth Systems

Crosscutting Concept (CCC): Stability and Change

Phenomenon: Urban Heat Islands and Local Climate Effects

Base your answers to questions 1 through 9 on the information below and on your knowledge of Earth and space sciences.

Urban Heat Islands (UHIs)

In large cities, temperatures tend to be higher than in surrounding rural areas. This is known as the urban heat island (UHI) effect. It occurs because pavement, buildings, and other artificial surfaces absorb and retain more heat than vegetation-covered land.

Scientists measured summer temperatures in different parts of a metropolitan area.

Average Surface Temperature by Land Type—July

Location Type	Average Daytime Temperature (°C)	Average Nighttime Temperature (°C)
Downtown (dense buildings)	37	29
Suburban (mixed land use)	33	24
Park (forested area)	28	20

1. Which conclusion is best supported by the data in the table?

(1) Parks retain more heat than developed areas.

(2) Temperature differences are greater at night than during the day.

(3) Urban surfaces lead to warmer conditions, especially at night.

(4) Temperature is unrelated to land use or surface material.

2. Using the data, explain how human development contributes to the urban heat island effect.

____

____

____

____

Modeling Urban Surfaces and Heat Absorption

Heat Absorption of Urban vs. Vegetated Sufaces

A. Urban Area

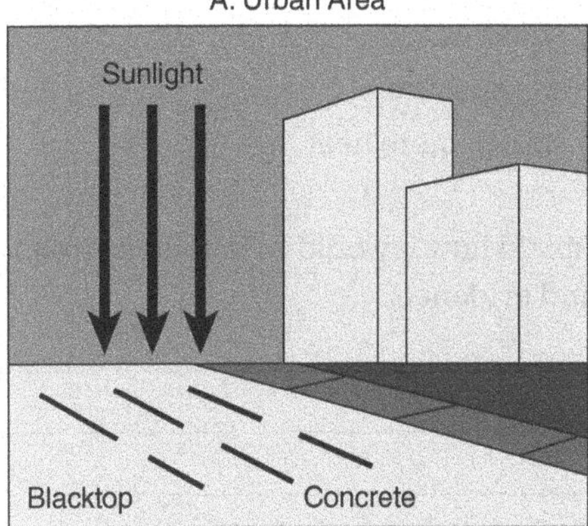

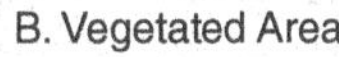

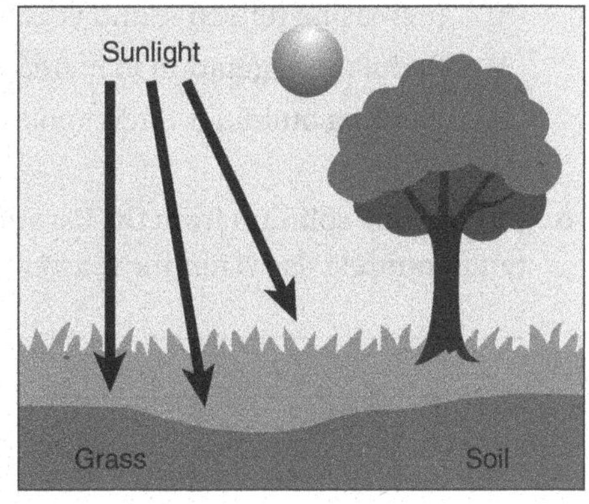

Urban materials absorb more solar energy than vegetation, leading to increased surface and air temperatures.

3. Which property of urban materials contributes most to higher temperatures?

(1) High albedo and strong reflection of sunlight

(2) Low density and high evaporation rates

(3) Low albedo and high heat absorption

(4) Increased water content and cooling capacity

4. Based on the model, describe how differences in surface materials affect temperature in urban and natural environments.

____

____

____

____

Designing Solutions: Cooling the City

City planners are testing methods to reduce urban heat island (UHI) effects. Some proposed solutions include:

- Installing green roofs (rooftop gardens)
- Replacing blacktop with light-colored, reflective materials
- Increasing tree planting and shaded areas

5. Which of the following strategies would most effectively reduce urban heat island effects?
   (1) Expanding roadways to allow better airflow
   (2) Increasing reflective and vegetated surfaces
   (3) Reducing vegetation to increase wind exposure
   (4) Painting buildings darker colors to absorb more heat

6. Choose one solution from the list and explain how it would help reduce urban temperatures. Use scientific reasoning and evidence.

____

____

____

____

Climate and Long-Term Impact

Scientists are concerned that urban heat islands (UHIs) may intensify climate change effects locally. They model changes in temperature and energy use for cooling buildings.

Projected Increase in Air Conditioning Use Due to UHI

Area Type	Increase in Energy Use by 2050 (%)
Urban	30
Suburban	20
Rural	5

7. Which effect is most likely caused by the urban heat island phenomenon?

(1) Decreased energy demand due to passive cooling
(2) Equal warming in all geographic areas
(3) Increased energy use for air conditioning in urban zones
(4) More heat in rural areas due to tree cover

8. Use the table to explain how UHIs affect energy consumption and contribute to climate feedback loops.

____

____

____

____

9. Claim: "Human activity alters local environments and contributes to broader climate impacts."

Using examples from the cluster, construct a well-supported argument to explain how local urban development can impact Earth's systems.

____

____

____

____

Answer Explanations

1. **(3)** The data show that urban areas (e.g., downtown) remain significantly warmer, especially at night. Artificial surfaces like concrete retain heat after sunset, while parks cool down more quickly. This supports the conclusion that urban surfaces lead to warmer nighttime conditions.
2. **Sample Response:** Human development replaces natural land with materials like asphalt and concrete, which absorb more heat during the day and release it slowly at night. This results in higher temperatures in urban areas compared with vegetated areas. The more densely developed a location is, the higher the surface and air temperature. This is the urban heat island effect.
3. **(3)** Urban surfaces like blacktop and concrete have low albedo, meaning they absorb a large amount of incoming solar radiation rather than reflecting it. This leads to higher temperatures. In contrast, surfaces with high albedo reflect more sunlight and stay cooler.
4. **Sample Response:** The model shows that dark, dense materials in urban areas (Panel A) absorb and retain more solar energy, resulting in higher surface temperatures. In vegetated areas (Panel B), plants and soil reflect more sunlight and also cool the environment through transpiration. These differences in surface properties cause urban areas to be hotter than natural landscapes.
5. **(2)** Increasing reflective surfaces and planting vegetation reduces the amount of heat absorbed and provides cooling through shade and evapotranspiration. These are proven methods for mitigating urban heat island effects and are supported by current climate adaptation strategies.
6. **Sample Response:** Planting trees provides shade and promotes evapotranspiration, which cools the surrounding air. Trees also reduce the amount of solar energy absorbed by paved surfaces. This helps lower temperatures in city areas, improving comfort and reducing energy used for cooling.
7. **(3)** The data in the table show that urban areas are projected to see the greatest increase in energy demand—up to 30%—because of the need for more air conditioning. This increase is directly related to higher temperatures caused by the urban heat island effect.

8. **Sample Response:** The data in the table show that urban areas are projected to see the greatest increase in energy demand—up to 30%—because of the need for more air conditioning. This increase is directly related to higher temperatures caused by the urban heat island effect.
9. **Sample Response:** Urban development increases surface temperatures by replacing vegetation with materials that absorb heat. This affects not only the local climate but also energy use and emissions. For example, warmer cities use more air conditioning, increasing greenhouse gas emissions and contributing to climate change. These examples show how local human activity can influence larger Earth systems and climate patterns.

Topic 5: History of Earth

Science and Engineering Practice (SEP): Interpreting and Analyzing Data

Disciplinary Core Idea (DCI): LS4.C: ESS1.C: The History of Planet Earth

Crosscutting Concept (CCC): Scale, Proportion, and Quantity

Phenomenon: Rock Layers and Fossils Reveal Earth's Long History

Base your answers to questions 1 through 9 on the information below and on your knowledge of Earth and space sciences.

Fossils in Rock Layers

Geologists use rock strata and fossils to understand Earth's history. Fossils are found in sedimentary rock layers and provide evidence of how environments and life forms have changed over time. The principle of superposition states that in undisturbed rock layers, the oldest layers are at the bottom and the youngest are at the top.

Fossil Distribution in Rock Layers at Site *A*

Rock Layer	Fossils Found
Layer 1 (top)	Leaf impressions, small mammals
Layer 2	Dinosaur bones, fern fossils
Layer 3	Marine shellfish, trilobites

1. What conclusion is best supported by the fossil evidence in the table?
   (1) Marine organisms lived before land organisms in this location.
   (2) Ferns evolved after mammals.
   (3) Dinosaurs and trilobites lived during the same time period.
   (4) Land environments existed before marine environments.

2. Using the fossil record in the table, explain how this site provides evidence of environmental changes over time.

____

____

____

____

Model of Rock Layers and an Intrusion

Rock Layers and Igneous Intrusion

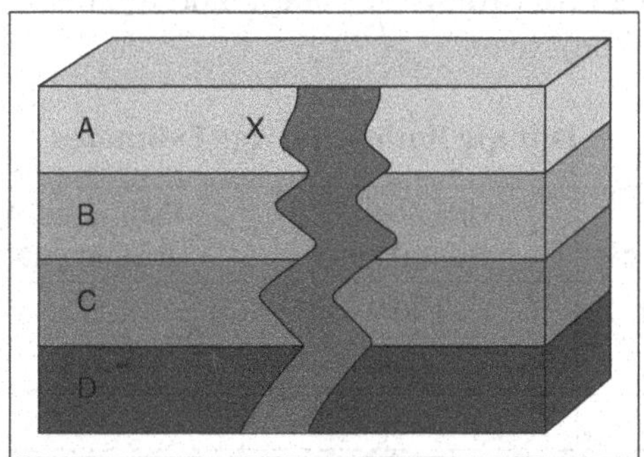

The cross section shows undisturbed sedimentary rock layers, an igneous intrusion, and a fault. This model helps determine the relative age of geologic features.

3. Which feature is the youngest in the figure?

(1) Layer *A*
(2) The fault
(3) The intrusion (*X*)
(4) Layer *D*

4. Based on the model, explain how geologists determine the relative ages of rock layers, faults, and intrusions.

____

____

____

____

Radiometric Dating

Radiometric dating allows scientists to assign actual ages to rocks and fossils using isotopes like uranium-238 or carbon-14. For example, uranium-238 decays to lead-206 with a half-life of 4.5 billion years.

Isotope Ratios and Age Estimates

Sample	% Uranium-238 Remaining	Estimated Age (billion years)
A	100%	0
B	50%	4.5
C	25%	9.0

5. What is the relationship between the amount of uranium-238 remaining and the estimated age of the sample?

(1) As uranium increases, age increases.
(2) As uranium decreases, age increases.
(3) As uranium decreases, age decreases.
(4) There is no relationship.

6. Using the table, explain how radioactive decay helps scientists estimate the age of Earth's oldest rocks.

____

____

____

____

Geologic Time and Extinction Events

Earth's 4.6-billion-year history is marked by significant biological and geological events, including mass extinctions, tectonic activity, and major climate shifts.

7. Which statement best describes the scale of geologic time?

(1) Most of Earth's major changes happened in the last 10,000 years.

(2) Earth's history is too short to be divided into eras or periods.

(3) Earth's history spans billions of years with life appearing relatively recently.

(4) The rock record contains no information about past extinctions.

8. Describe how both rock layers and fossil records provide evidence for major changes in Earth's history, including extinction events.

____

____

____

____

9. Claim: "Rock layers and fossil records provide evidence for major changes in Earth's history, including extinction events."

Using examples from the cluster, construct a well-supported argument to explain how scientists use rock and fossil evidence to reconstruct Earth's history.

____

____

____

____

Answer Explanations

1. **(1)** According to the principle of superposition, Layer 3 is the oldest. The presence of marine fossils like trilobites in that layer indicates that the region was once underwater. Higher up in the sequence, fern fossils and dinosaur bones in Layer 2 suggest a shift to a land environment with different life forms. Finally, Layer 1 contains fossils of small mammals and leaves, indicating a more modern terrestrial environment. This progression supports the conclusion that marine organisms lived before land organisms in this location.
2. **Sample Response:** The fossil record in the table shows a transition from marine organisms (trilobites, shellfish) in Layer 3 to dinosaurs and ferns in Layer 2 and then to small mammals and leaves in Layer 1. This sequence indicates a shift from an ocean environment to a terrestrial ecosystem. These fossils provide evidence of both environmental change (from sea to land) and biological evolution (from simple marine life to complex land-dwelling organisms). This demonstrates how rock layers preserve a timeline of Earth's changing surface conditions and ecosystems.
3. **(1)** The fault is the youngest feature. It cuts across Layers *B* through *D* but not Layer *A*. This means Layer *A* was deposited after the faulting occurred. According to the law of crosscutting relationships, a fault or intrusion is younger than the layers it cuts through, and if a layer is not cut, it is younger than the fault. Therefore, the fault must have occurred after Layers *B, C,* and *D* were deposited but before Layer *A* formed.
4. **Sample Response:** Geologists use the principle of superposition to determine that deeper sedimentary layers are older than those above them. They also apply the principle of crosscutting relationships: any fault or igneous intrusion that cuts through existing layers is younger than those layers. In the figure, the igneous intrusion (*X*) is younger than Layers *A, B,* and *C* because it cuts through them but older than Layer *D*, which it does not reach. The fault offsets Layers *B* through *D*, making it younger than those layers but older than Layer *A*. This process allows geologists to determine the relative order of events even without absolute dates.
5. **(2)** As shown in the table, samples with less uranium-238 remaining are older. Uranium-238 decays over time into lead-206. A sample with 50% of the original uranium-238 remaining has undergone 1 half-life (4.5 billion years). If only 25%

remains, 2 half-lives (9.0 billion years) have passed. Therefore, the less uranium remaining, the greater the sample's age. This is the basis of radiometric dating.

6. **Sample Response:** Radioactive isotopes like uranium-238 decay at a predictable rate. Scientists use the proportion of parent isotope (uranium-238) to daughter isotope (lead-206) to determine how many half-lives have passed. In the table, Sample *C* has only 25% of its original uranium left, meaning 2 half-lives (9 billion years) have passed. By measuring isotope ratios in ancient rocks, scientists can calculate their ages and determine that Earth is approximately 4.6 billion years old.

7. **(3)** Earth's history spans about 4.6 billion years. Complex life has existed for only the last few hundred million years, a small fraction of that time. The geologic time scale breaks Earth's history into eons, eras, and periods based on major changes such as the emergence of life, mass extinctions, and tectonic activity. This large scale allows scientists to study Earth's gradual and sudden changes over vast time frames.

8. **Sample Response:** Rock layers show changes in composition, thickness, and types of sediments, which can be tied to ancient environments (like oceans, deserts, or volcanic eruptions). Fossils within those layers help identify which organisms lived during specific time periods. When certain fossils suddenly disappear from the rock record, it suggests an extinction event. For example, the disappearance of dinosaur fossils between certain layers marks the Cretaceous-Paleogene mass extinction. Together, rock and fossil records provide detailed evidence of Earth's dynamic history, including life's evolution and catastrophic events.

9. **Sample Response:** Geologists and paleontologists study rock layers and fossil evidence to reconstruct Earth's past. At Site *A*, the progression from marine fossils in deeper layers to terrestrial fossils in shallower layers shows environmental changes over time. The cross section at Site *B* reveals how intrusions and faults disrupt older rock layers, helping scientists determine the sequence of geological events. Radiometric dating provides absolute ages for these events, confirming that Earth has changed significantly over billions of years. These tools and techniques demonstrate that both Earth's surface and its living organisms have experienced dramatic transformations over time.

Topic 6: Natural Resources and Sustainability

Science and Engineering Practice (SEP): Designing Solutions

Disciplinary Core Idea (DCI): ESS3.A/B: Natural Resources/Human Impacts

Crosscutting Concept (CCC): Energy and Matter

Phenomenon: Freshwater Scarcity and Groundwater Depletion

Base your answers to questions 1 through 9 on the information below and on your knowledge of Earth and space sciences.

Water Under Pressure: The Problem of Groundwater Depletion

In many agricultural regions, particularly in the western United States, water is drawn from underground aquifers to meet the demands of irrigation, drinking water, and industry. However, groundwater is being pumped faster than it can be naturally replenished, resulting in declining water tables and long-term sustainability concerns.

Changes in Groundwater Levels in Central Valley, California

Year	Average Groundwater Level (meters below surface)
2000	15.2
2005	17.4
2010	19.8
2015	22.6
2020	25.9

1. Which conclusion is best supported by the data in the table?
   (1) Groundwater levels rise as more wells are drilled.
   (2) The water table is becoming shallower over time.
   (3) Groundwater levels are steadily decreasing over time.
   (4) Water use does not affect aquifer depth.

2. Using evidence from the table, describe how human activity is affecting groundwater availability in the Central Valley.

____

____

____

____

Stimulus: Cross Section of an Aquifer System

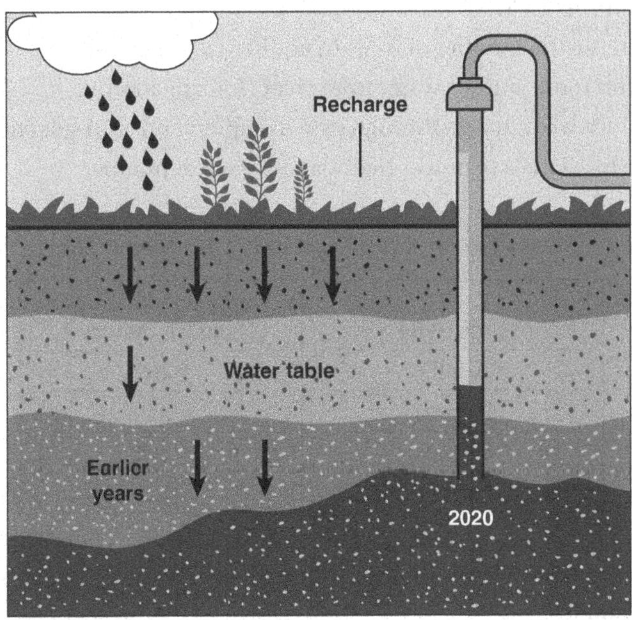

3. Which feature in the model best represents groundwater recharge?
   (1) Arrows pointing downward from rainfall into soil
   (2) Water flowing from rivers into the ocean
   (3) Pumps extracting water from deep wells
   (4) Evaporation from irrigation canals

4. Explain how overpumping groundwater affects the water table and energy use in farming communities.

____

____

____

____

5. Which of the following is the best example of a sustainable solution for preserving groundwater resources?

(1) Increase the number of wells in aquifer-rich regions.
(2) Divert more surface water from rivers for irrigation.
(3) Reduce water usage through drip irrigation and crop selection.
(4) Use fossil fuels to power high-capacity water pumps.

Water Use by Crop Type in the Southwest U.S.

Crop Type	Water Use (liters per square meter)
Alfalfa	1,200
Cotton	950
Grapes	520
Lettuce	420

6. Based on the table, which strategy would most effectively reduce agricultural groundwater use?

(1) Replace lettuce fields with cotton.
(2) Increase alfalfa production during droughts.
(3) Grow more grapes and lettuce instead of high-use crops.
(4) Prioritize crops with the highest water demand.

7. Using the data, describe how crop choice influences groundwater sustainability.

____

____

____

____

8. Which process best illustrates the cycling of matter and the flow of energy in a groundwater system?

(1) Drilling new wells to access deeper water

(2) Planting trees in dry environments

(3) Water infiltrating the ground and recharging aquifers

(4) Using pesticides that reach the water table

9. Claim: "Human water use must be balanced with natural recharge to ensure groundwater sustainability."

Using examples from the cluster, construct a well-supported argument to explain why managing water use is essential for long-term aquifer health.

____

____

____

____

Answer Explanations

1. **(3)** Groundwater depth increases from 15.2 to 25.9 meters over 20 years, indicating that water is being removed faster than it is being replenished. This pattern shows a steady decline in groundwater availability.
2. **Sample Response:** The increasing depth of groundwater shown in the table reflects long-term overuse. As more water is pumped from the aquifer, the water table drops, requiring deeper wells and threatening future access to clean water.
3. **(1)** Recharge is the process where water from precipitation infiltrates through soil into the aquifer. In the model, this is represented by arrows showing rainwater soaking into the ground—the key to replenishing depleted groundwater.
4. **Sample Response:** When groundwater is overpumped, the water table drops and wells must be drilled deeper. Pumping from greater depths requires more energy, increasing costs for farmers. This strain can reduce agricultural productivity and harm local economies.
5. **(3)** Reducing water use through methods like drip irrigation and selecting drought-tolerant crops directly addresses the cause of groundwater depletion, making this the most sustainable and effective strategy listed.
6. **(3)** Alfalfa and cotton consume far more water than grapes or lettuce. Shifting to crops that require less water would reduce irrigation demand and preserve aquifer levels.
7. **Sample Response:** By selecting crops that require less water—like lettuce and grapes—farmers can significantly lower groundwater usage. The table shows that alfalfa uses nearly three times more water per square meter than lettuce, so changing crop types can make a big difference.
8. **(3)** The infiltration of water into the ground is part of the hydrologic cycle, showing both matter cycling and energy flow. This process helps restore groundwater and sustains ecosystems and agriculture.
9. **Sample Response:** The data show that groundwater is being depleted faster than it is being replenished. Sustainable practices—like efficient irrigation and changing crop types—help slow the drop in water levels. Recharge from rainfall takes time, so water use must be carefully managed to maintain a balance and prevent long-term scarcity.

Topic 7: Engineering, Technology, and Applications of Science

Science and Engineering Practice (SEP): Designing Solutions

Disciplinary Core Idea (DCI): ETS1.B: Developing Possible Solutions

Crosscutting Concept (CCC): Stability and Change

Phenomenon: Managing Coastal Erosion with Engineering Design

Base your answers to questions 1 through 9 on the information below and on your knowledge of Earth and space sciences.

Coastal Erosion at Oceanview Beach

Oceanview Beach is losing shoreline each year due to wave action, storm surges, and rising sea levels. Engineers are testing different methods to reduce erosion and protect local homes and ecosystems.

Comparison of Coastal Protection Strategies

Strategy	Description	Cost Estimate	Effectiveness Rating (1–5)
Seawall	Concrete barrier parallel to shore	High	4
Beach nourishment	Adding sand to replace eroded areas	Medium	3
Vegetation planting	Using native plants to stabilize dunes	Low	2
Offshore breakwater	Submerged structure to reduce wave energy	High	5

1. Which strategy is rated most effective at protecting the coastline from erosion?
   (1) Seawall
   (2) Beach nourishment
   (3) Vegetation planting
   (4) Offshore breakwater

2. Using the data in the table, explain how engineering solutions are evaluated based on both effectiveness and cost.

____

____

____

____

Modeling Wave Reduction and Breakwater Placement

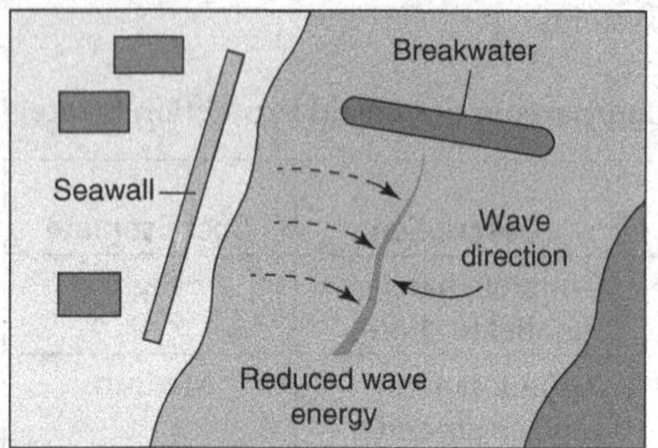

Breakwaters are placed offshore to reduce wave energy before it reaches the shore, while seawalls protect against erosion from direct wave impact.

3. What is the primary function of a breakwater?
   (1) Absorb heat from ocean water
   (2) Slow wave energy before it reaches the shore
   (3) Increase longshore drift to reshape the beach
   (4) Reflect solar radiation into the ocean

4. Based on the diagram, explain how the placement of breakwaters helps stabilize shorelines and protect infrastructure.

____

____

____

____

Balancing Engineering and the Environment

Although structures like seawalls and breakwaters are effective, they can also disrupt natural habitats or alter sediment movement. Engineers must consider long-term impacts when designing coastal protection systems.

5. Which of the following represents a trade-off in coastal engineering?
   (1) Using only natural materials regardless of cost
   (2) Allowing the shoreline to erode naturally
   (3) Preventing erosion while potentially harming ecosystems
   (4) Building taller seawalls to improve views

6. Describe how engineers balance environmental concerns with human needs when designing coastal protection systems.

____

____

____

____

Evaluating Solutions Through Modeling and Testing

Engineering design includes multiple steps: defining a problem, developing and testing possible solutions, and refining designs based on feedback and results.

7. Which step in the engineering design process involves using scale models or simulations to predict performance?
   (1) Defining the problem
   (2) Identifying materials
   (3) Testing solutions
   (4) Implementing laws

8. Give an example of how an engineer could use data from a coastal monitoring project to improve a shoreline protection design.

9. Claim: "Engineering designs can reduce environmental impacts while solving real-world problems."

   Using examples from the cluster, construct a well-supported argument to explain how engineering is used to protect Earth's systems.

Answer Explanations

1. **(4)** The offshore breakwater is rated as the most effective strategy, with a score of 5 in the "Effectiveness Rating" column of the table. Breakwaters are submerged or partially submerged barriers placed offshore that significantly reduce the energy of incoming waves before they reach the shore. By absorbing and deflecting wave force, breakwaters help prevent erosion and protect coastlines more efficiently than other listed methods. Although both the breakwater and the seawall are marked "High" in cost, the breakwater's higher effectiveness rating makes it the best-performing solution overall.
2. **Sample Response:** Engineering design involves evaluating potential solutions based on multiple criteria, including effectiveness, cost, durability, and environmental impact. The table presents four different erosion control strategies, each with trade-offs. For example, vegetation planting is the least expensive but also the least effective (rated 2). In contrast, offshore breakwaters are highly effective (rated 5) but require a large investment due to material and construction demands. Engineers must weigh these factors and consider local conditions—such as storm frequency, economic resources, and environmental sensitivity—before selecting a solution. Ultimately, the goal is to implement a solution that is both sustainable and appropriate for the region's specific needs.
3. **(2)** The primary function of a breakwater is to reduce wave energy before it reaches the shore. This function is essential in managing erosion and preventing property damage along coastlines. By disrupting the force of waves offshore, breakwaters protect beaches, dunes, and coastal infrastructure, allowing sand to accumulate and decreasing the rate of shoreline retreat. Breakwaters are particularly effective in areas exposed to high-energy wave environments, such as open coastlines and hurricane-prone zones.
4. **Sample Response:** The figure shows the strategic placement of a breakwater parallel to the coastline, positioned far enough from the shore to intercept wave energy. Waves lose much of their momentum as they encounter the breakwater, which creates a calm zone between the breakwater and the beach. This reduced wave force limits the scouring effect of waves on sand and dunes. It also prevents the direct erosion of shorelines and artificially made structures. Additionally, calmer waters behind the breakwater support sediment deposition, which helps naturally replenish the beach. This protective mechanism is crucial in preserving ecosystems and maintaining safe conditions for coastal communities.

5. **(3)** This question targets the concept of engineering trade-offs, where solving one problem can introduce new challenges. Building hard structures like seawalls or breakwaters may effectively prevent erosion and protect infrastructure. However, they often alter natural wave patterns and sediment movement, potentially harming coastal ecosystems such as wetlands, coral reefs, or nesting grounds for marine life. This is a classic example of a conflict between human development and environmental preservation and is one of the primary ethical and ecological considerations in environmental engineering.
6. **Sample Response:** Coastal engineers must find ways to balance human safety, property protection, and environmental sustainability. This includes using environmentally conscious materials, blending natural and artificial strategies, and minimizing habitat disruption. For instance, combining dune restoration with elevated walkways and limited-use seawalls may protect infrastructure while preserving native vegetation. Engineers also work closely with environmental scientists and urban planners to assess the long-term impacts of their designs and ensure they do not degrade ecosystems. Environmental impact assessments, local stakeholder feedback, and real-time monitoring are integrated into the engineering process to create solutions that are adaptive and socially responsible.
7. **(3)** The testing phase of the engineering design process involves creating models, simulations, or prototypes to evaluate a design before full-scale implementation. In the case of coastal protection, engineers may use computer simulations to model wave impact on different barrier placements, or they may test scale models in wave tanks to observe real-time responses. Testing allows engineers to refine designs, identify weaknesses, and make data-driven improvements. This step is essential in ensuring that the final product is both safe and effective.
8. **Sample Response:** Engineers can use data from coastal monitoring projects, including GPS shoreline mapping, tide gauges, drone imagery, and buoy-based wave sensors. For example, if data show that beach erosion increases after storms, engineers may analyze whether existing protections are sufficient or whether modifications—such as extending a breakwater or reinforcing a seawall—are necessary. Monitoring can also inform timing and methods for beach nourishment, helping engineers select areas that will retain sand most efficiently. These feedback loops between monitoring and design are essential for adapting to dynamic coastal systems, especially as sea levels rise and weather patterns become more extreme.

9. **Sample Response:** Engineering plays a vital role in managing environmental challenges in a responsible and innovative way. At Oceanview Beach, the engineering team evaluated four strategies for mitigating erosion. Through data analysis, modeling, and field observation, they determined that the offshore breakwater provided the highest effectiveness. However, they also considered cost, ecological effects, and community input in their decision. By designing a breakwater that minimizes ecological disruption while maximizing shoreline protection, the team demonstrated how engineering can solve real-world problems and enhance the stability of natural systems. This approach embodies the three-dimensional model of science learning: applying scientific knowledge (DCI), practice (SEP), and conceptual understanding (CCC) to protect Earth's systems and improve human resilience.

Chapter 7

Regents Practice Exams

Regents Practice Exam June 2025

Directions: **For each multiple-choice question, record in the space provided the number of the choice that best completes the statement or answers the question. For questions requiring a written response, write your answers in the spaces provided.**

Base your answers to questions 1 through 5 on the information below and on your knowledge of Earth and Space Sciences. Some questions may require the use of the ***2024 Edition Reference Tables for Earth and Space Sciences.*** Available at *https://www.nysed.gov/sites/default/files/programs/state-assessment/earth-science-reference-tables-english-2011.pdf*

Our Sun—A Star

The Hertzsprung-Russell (H-R) diagram was developed from star charts by two scientists in different countries independently of each other in 1911. It classified stars based on their surface temperatures, observable color, and magnitude. Absolute magnitude is a measurement of how bright a star would appear if all stars are the same distance from Earth. The brighter the star, the lower the absolute magnitude value.

1. Our Sun is classified as a spectral class G star with a surface temperature between 5000 to 6000 K and an absolute magnitude of about five. Based on this information, complete the H-R diagram model by placing **one X** to indicate where the Sun is located. Also, identify the relative temperature and relative absolute magnitude of the Sun as it transitions to a red giant. [1]

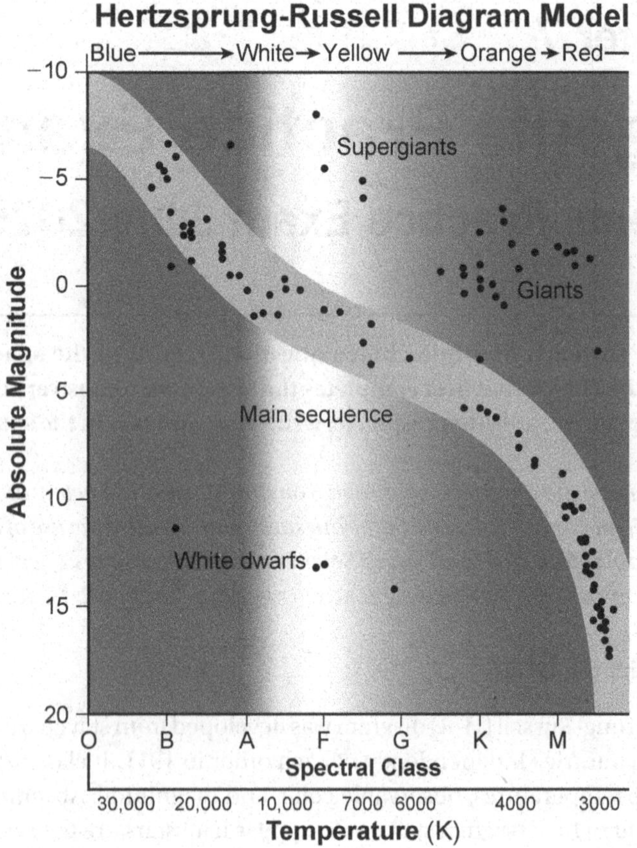

Change in relative temperature: ___________________________

Change in relative absolute magnitude: ___________________________

The model below shows the layers of the Sun and information about some features of each layer.

Model of Sun's Layers

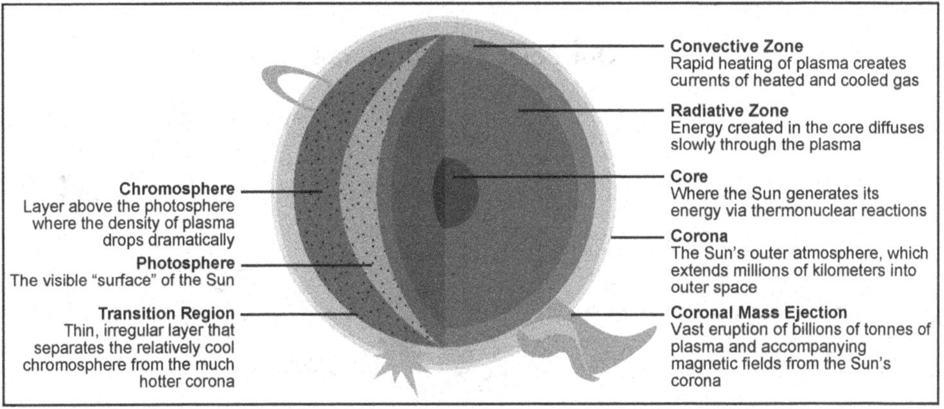

(Not drawn to scale)

2. Based on the information in the model, which list of five of the Sun's layers are in the correct sequence to allow energy generated by fusion to eventually reach the Sun's surface as radiation?

(1) Core → chromosphere → photosphere → transition region → corona

(2) Core → radiative zone → transition region → photosphere → corona

(3) Core → photosphere → corona → transition region → chromosphere

(4) Core → radiative zone → convective zone → photosphere → chromosphere

2 ______

The model below represents the orbits of celestial bodies around our Sun. The inset model shows some information about the region of the solar system inside the orbit of Jupiter.

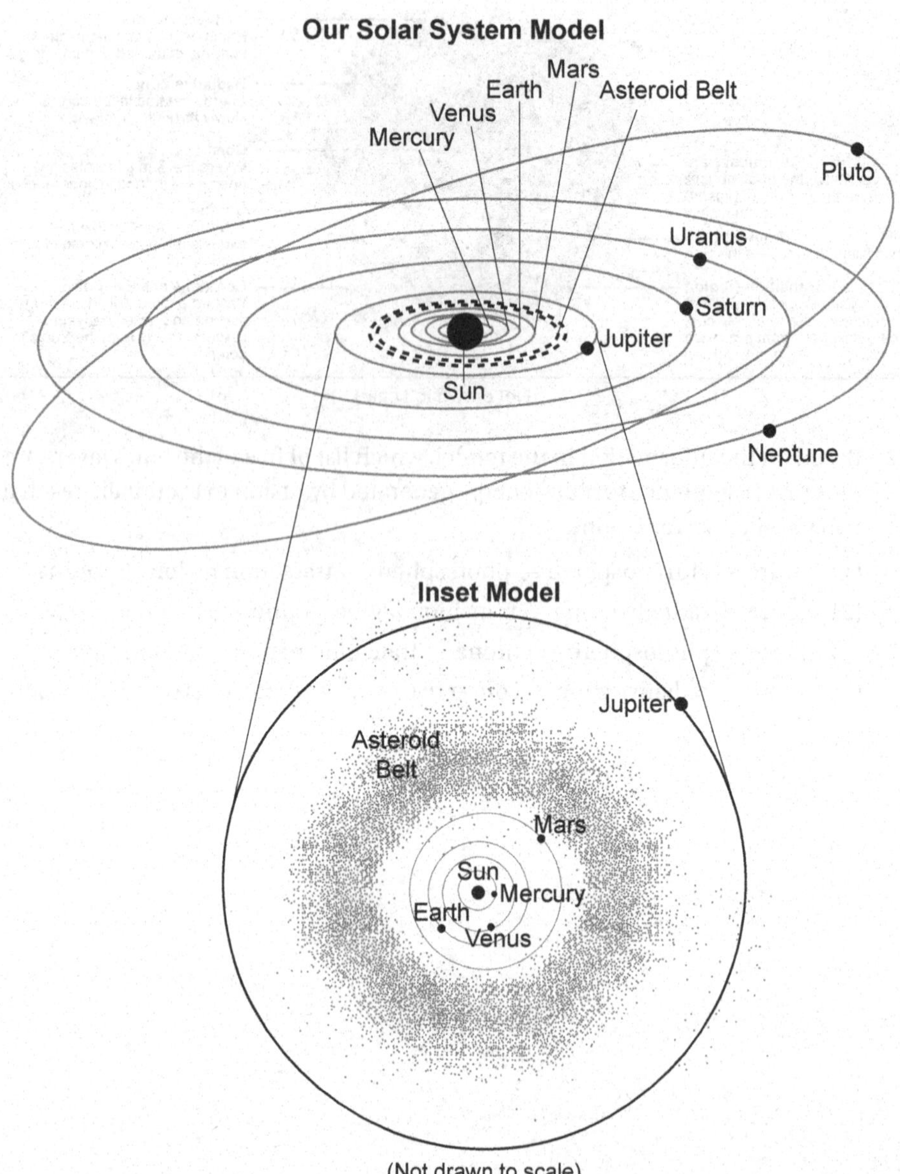

(Not drawn to scale)

3. The table below shows the eccentricity of the orbits of Mercury and Venus around the Sun.

Planet	Eccentricity
Mercury	0.206
Venus	0.007

Place a check mark (✓) in the boxes to indicate the **two** statements that are correctly predicted by Kepler's Laws. [1]

☐ Venus orbits the Sun at a constant speed.

☐ Mercury travels faster in its orbit when it is closer to the Sun.

☐ Venus's orbit is less elliptical than Mercury's orbit.

☐ The orbital speeds of both planets are affected by their masses.

☐ Unlike Venus, the eccentricity of Mercury's orbit prevents Mercury from having a moon.

4. Based on the *Our Solar System Model,* if a new planet was identified that orbited the Sun at an average orbital distance greater than that of Mercury but less than that of Venus, the average speed of this planet would be

(1) greater than the average speed of Mercury, but less than the average speed of Venus.

(2) less than the average speed of Mercury, but greater than the average speed of Venus.

(3) greater than the average speed of Venus, but less than the average speed of Earth.

(4) less than the average speed of Venus, but greater than the average speed of Earth.

4 ______

An observer on Earth sees phases of the Moon, but the Moon isn't the only solar system body to exhibit phases. Venus also has observable phases as viewed from Earth. Venus's orbit around the Sun is approximately 225 Earth days.

The photograph below shows a Moon phase and Venus viewed with unaided eyes in the night sky. The inset box shows Venus observed using a telescope. Both the Moon and Venus are in crescent phase.

Observed Phases of the Moon and Venus

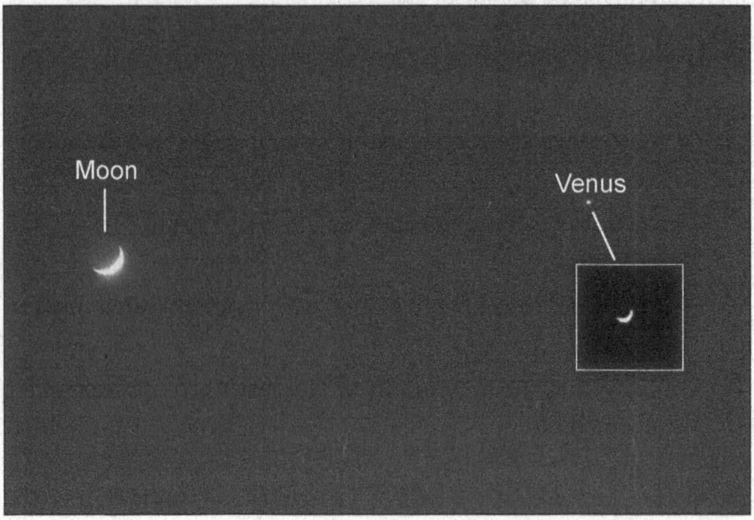

Moon

Venus

5. Using *Our Solar System Model*, construct an explanation for why an observer on Earth can see a cycle of phases for the planet Venus. In the spaces below, write the terms for choices *A*, *B*, and *C* that correctly complete the passage. [1]

Choices A:	**Choices B:**	**Choices C:**
• inside	• closer to	• Moon
• outside	• farther from	• Sun

Venus orbits the Sun, circling ___A___ Earth's orbit in about 225 Earth days. This means that Venus is sometimes ___B___ Earth, while at other times it is positioned on the other side of the ___C___ . It is this change in relative positions of Venus that causes an observer on Earth to see phases of Venus.

Choice *A*: ____________________

Choice *B*: ____________________

Choice *C*: ____________________

Base your answers to questions 6 through 10 on the information below and on your knowledge of Earth and space sciences. Some questions may require the use of the ***2024 Edition Reference Tables for Earth and Space Sciences****.*

Modeling Earth Systems to Understand Global Climate

The geologic record shows a long history of climate fluctuations as a result of many different factors. Climate scientists study models of Earth's motions, ocean currents, plate tectonic movement, and atmospheric composition to better understand energy flow into and out of Earth's systems.

Obliquity, the tilt of Earth's axis relative to the Sun, has a direct impact on Earth's climate. Earth's obliquity changes on a cycle that takes place over a period of 41,000 years. The obliquity is currently decreasing and will reach its minimum value of 22.1° in approximately 9800 years.

Maximum and Minimum Angles of Obliquity

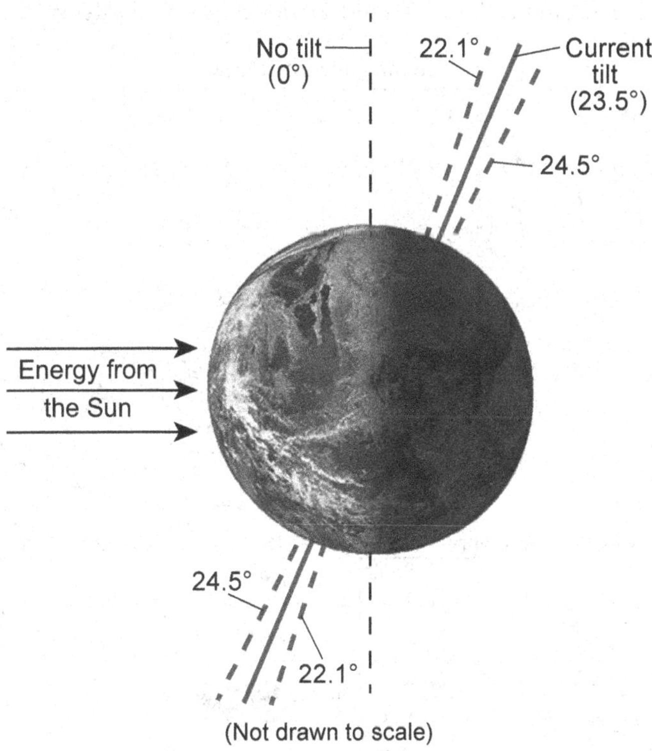

(Not drawn to scale)

6. During winter, which statement best describes the amount of energy Earth's northern hemisphere will receive and the impact on global ice formation when the obliquity is 22.1°, compared to Earth's present obliquity?

(1) Earth's northern hemisphere will receive less energy, and less ice will form in polar regions.
(2) Earth's northern hemisphere will receive less energy, and more ice will form in polar regions.
(3) Earth's northern hemisphere will receive more energy, and less ice will form in polar regions.
(4) Earth's northern hemisphere will receive more energy, and more ice will form in polar regions.

6 ______

Another factor climate scientists have identified that contributes to changes in energy flow in Earth systems is ocean current circulation.

The model, *Diagram 1,* and *Diagram 2* show some information about circulation patterns in ocean currents. Points *X* and *Y* are locations on Earth's surface.

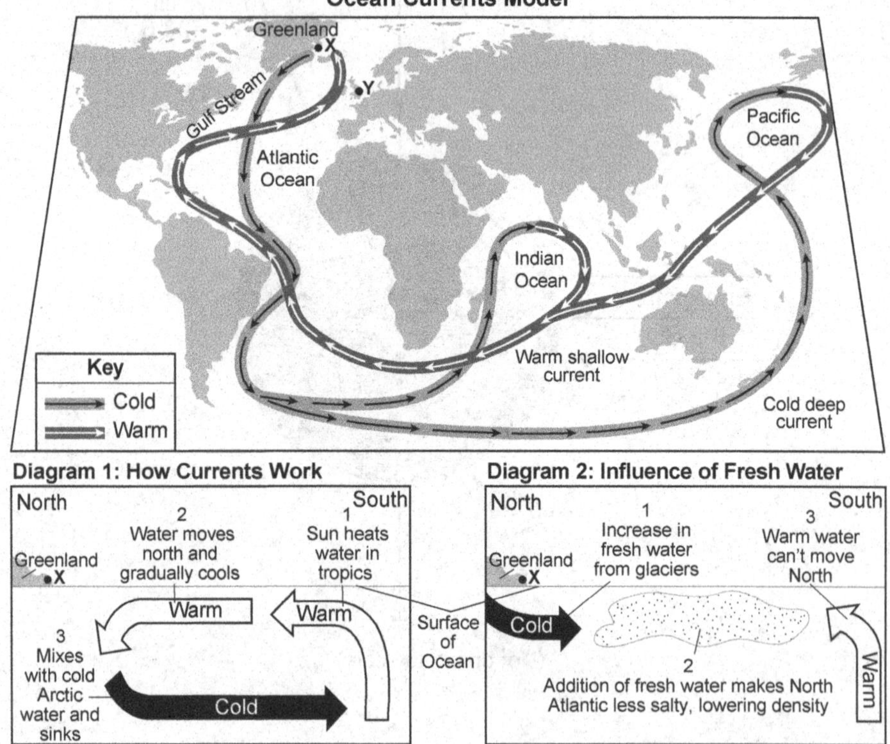

7. Which statement most accurately describes the influence of melting glaciers at *X* on the strength of Earth's ocean currents?

(1) More fresh water is added to the oceans, causing the currents to become deeper.
(2) More fresh water is added to the oceans, causing the currents to weaken.
(3) More fresh water is added to the oceans, causing the water in the currents to become denser.
(4) More fresh water is added to the oceans, causing the water in the currents to become warmer.

7 ______

8. Which statement most accurately describes the influence of present surface ocean currents on the climate at location *Y*?

(1) Location *Y* experiences warmer air temperatures with more precipitation.
(2) Location *Y* experiences warmer air temperatures with less precipitation.
(3) Location *Y* experiences cooler air temperatures with more precipitation.
(4) Location *Y* experiences cooler air temperatures with less precipitation.

8 ______

Since the Industrial Revolution, deposition of dark particles such as dust, dirt, and rock in glacial ice (contaminated snow) have caused glaciers to darken. This has led to feedbacks that have caused changes to other Earth systems.

Percent of Sunlight Reflected and Absorbed by Different Glacial Surfaces

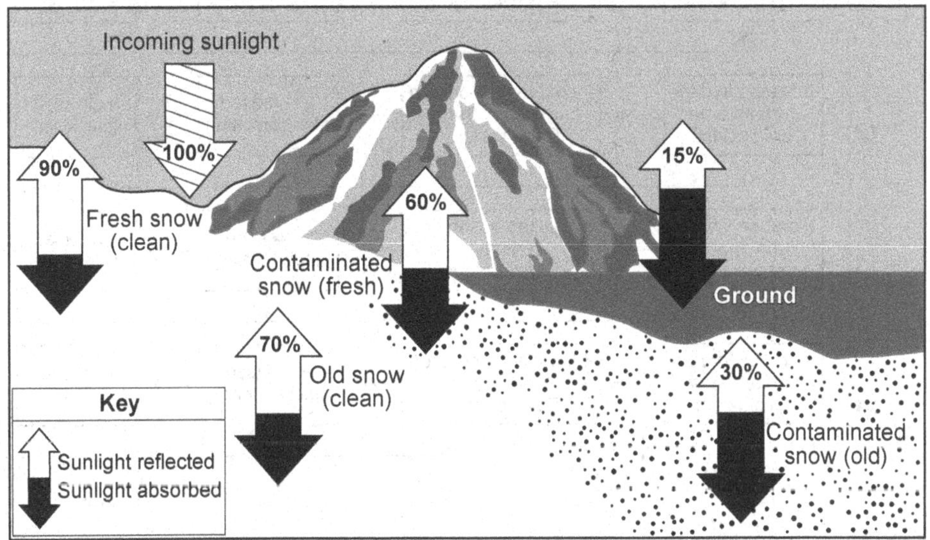

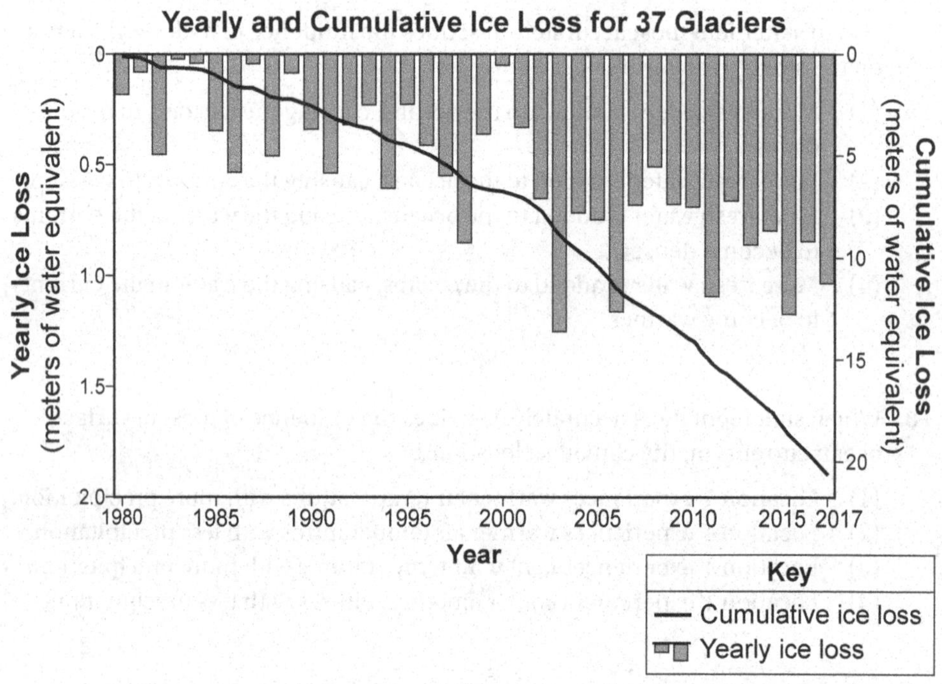

9. Which model correctly represents the feedbacks that occur when dark surfaces are exposed in glacial areas and cause changes to one or more Earth systems?

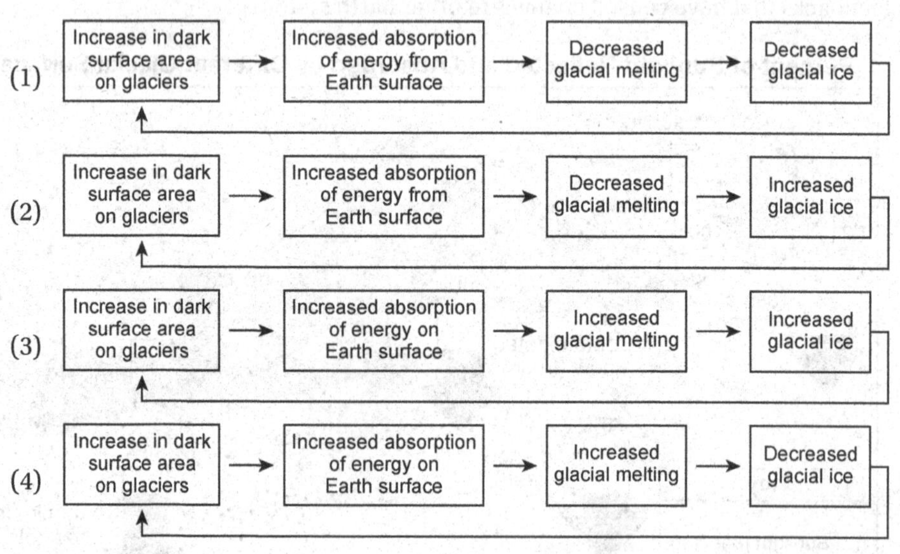

9 _____

Computer-based global climate models are helpful tools for collecting data on projected future climate conditions. These models use various scenarios, or possibilities, that assume different human-based decisions on how we address greenhouse gas emissions.

The graph below shows future greenhouse gas concentrations for four different greenhouse gas emission scenarios measured in parts per million (ppm).

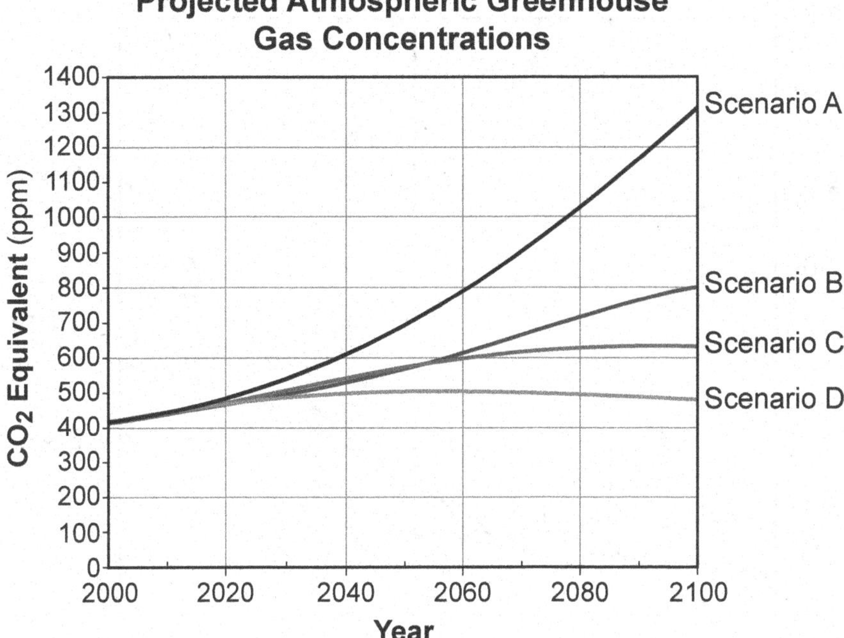

The graph below shows some information about global surface air temperature changes. These models are compared to the average global surface air temperature between the years 1986 to 2005, indicated as 0.0°C.

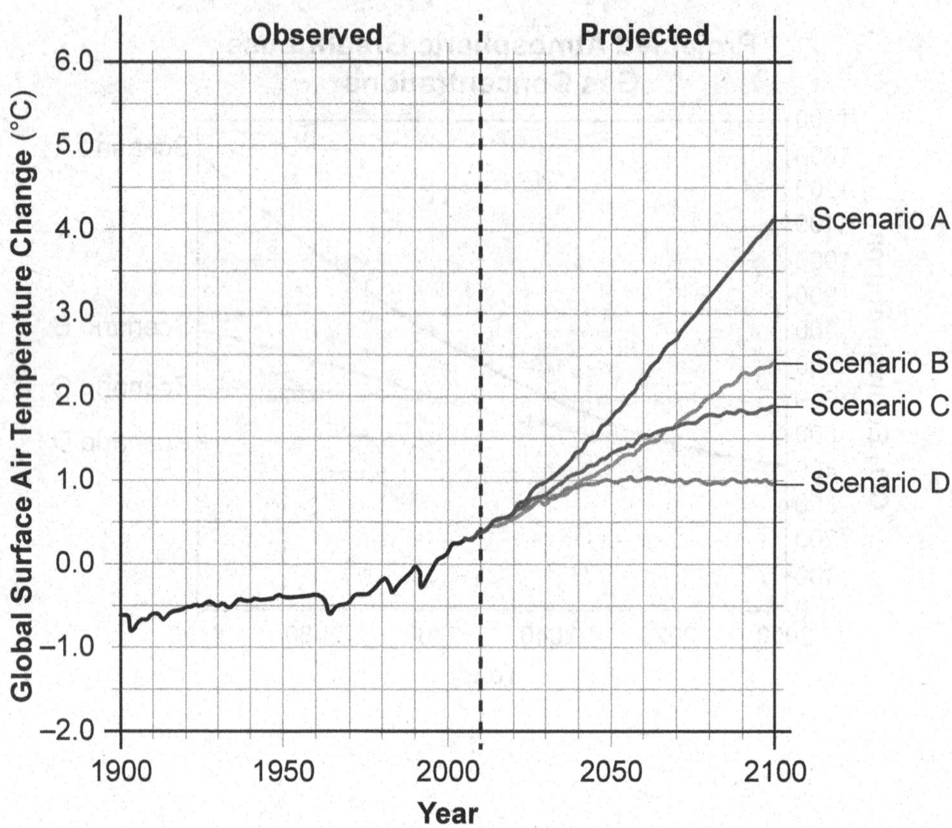

10. Identify the numerical values for the projected concentration of greenhouse gases (CO_2 equivalent) and for the approximate future change to global surface temperature for the year 2100 using scenario *B* emissions. [1]

Projected CO_2 equivalent in 2100: ______________________________ **ppm**

Projected global surface temperature change: ______________________ **°C**

*Base your answers to questions 11 through 15 on the information below and on your knowledge of Earth and space sciences. Some questions may require the use of the **2024 Edition Reference Tables for Earth and Space Sciences**.*

The Carbon Cycle

The global carbon cycle refers to the movement of the element carbon through different storage places, or reservoirs, on Earth. Carbon moves through these reservoirs at different rates. Most carbon near Earth's surface cycles fairly quickly. Carbon in the atmosphere recycles in about three to five years, while plants recycle carbon in about 50 years. The carbon found in soil and fossil reservoirs is recycled, on average, in about 3000 to 5000 years.

The carbon cycle has two parts. The "fast cycle" involves the biological processes of photosynthesis and decomposition. The "slow cycle" involves the time it takes for soil (inorganic) carbon to form from the weathering of rocks and soil.

Wildfire events contribute to the carbon cycle. In 2020, megafires released an estimated 107 million metric tons of carbon dioxide into the atmosphere—equivalent to the amount released by about 23 million cars.

The model below shows some information about Earth's systems and the carbon cycle.

Model of Effects of Forest Fires and Burning Fossil Fuels on the Carbon Cycle

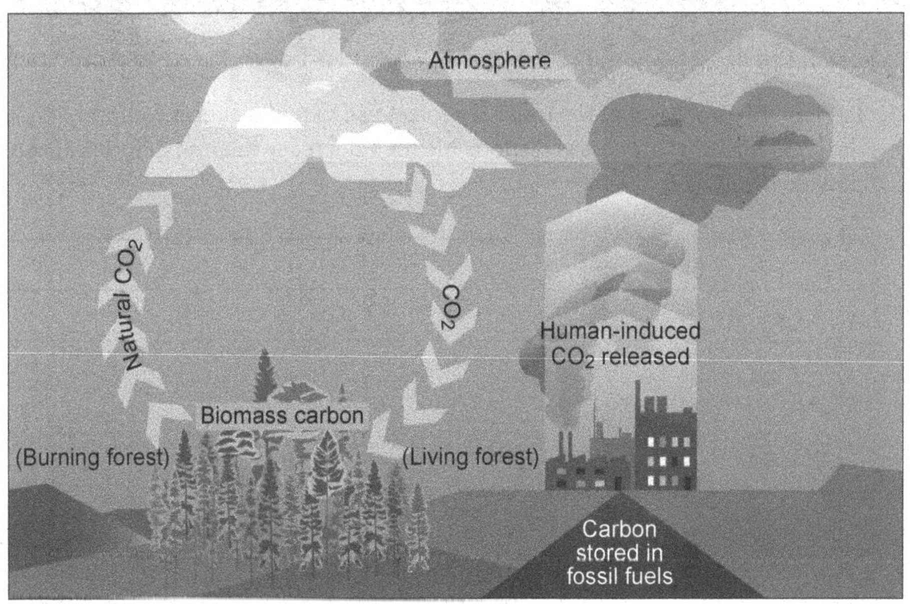

The model below represents how carbon levels change depending on what is happening with the trees in a forest.

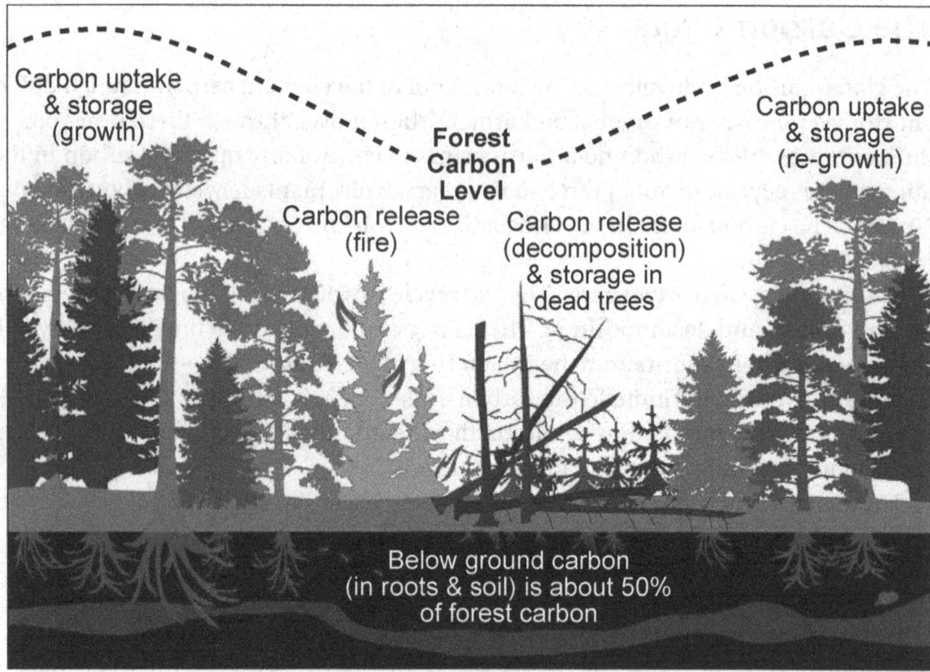

11. Explain the process that the trees in a forest use to make energy for food *and* describe how this process is responsible for a decrease in atmospheric carbon dioxide levels. [1]

____

____

____

The model below shows the movement of carbon (arrows) through the four Earth spheres. The numbers indicate the amount of carbon naturally added or removed from the spheres in gigatons (GT) per year. The numbers in bold indicate the amount of carbon added or removed by human activity. Numbers in parentheses ( ) are amounts of stored carbon.

Model of Fast Carbon Cycle

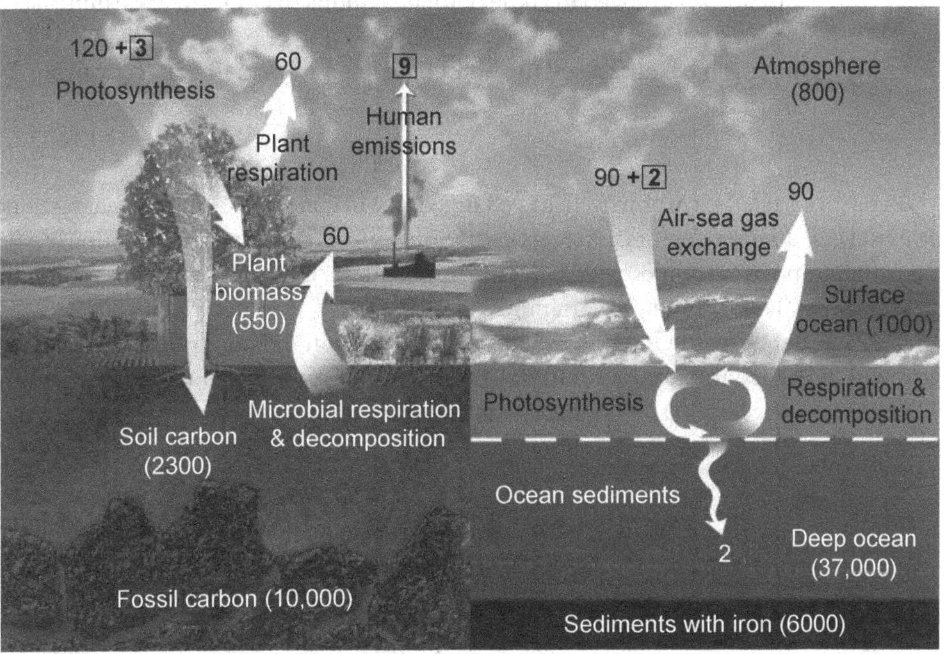

12. Which statement correctly identifies the quantitative cycling of carbon between two of Earth's spheres as a result of natural processes and human activities?

(1) Fossil carbon releases 2300 GT, while microbial respiration and decomposition absorb 60 GT.

(2) The deep ocean stores 36,000 more GT of carbon than is released by air-sea gas exchange.

(3) Human emissions add nine times more carbon to the atmosphere than plant respiration, which is the same amount released to the atmosphere by microbial respiration and decomposition.

(4) The amount of carbon that leaves the atmosphere and is absorbed by the ocean is 92 GT, which is the same amount that is released by the oceans back into the atmosphere and absorbed by ocean sediments.

12 ______

13. Which explanation describes how climate change from increased atmospheric carbon dioxide has influenced human activity?

(1) Humans have increased the replanting of trees in areas burned by wildfires in order to decrease the amount of local atmospheric carbon dioxide.
(2) Humans have increased the burning of fossil fuels in order to decrease the amount of carbon dioxide in the atmosphere.
(3) Humans have moved to cooler climate regions to adjust to a warming climate.
(4) Humans have decreased the number of dead trees in the forest by using them as fuel.

13 ______

In Earth's past, the carbon cycle has changed due to changes in climate that resulted from several different factors. Changes in the Sun's energy, the amount of marine organisms that remove carbon dioxide from the atmosphere, and uplift of major mountain chains have all contributed to variations in CO_2.

Ice-core data provide a record of atmospheric carbon dioxide (measured from air trapped in the ice in parts per million (ppm)) and of Antarctic surface temperature changes over the last 800,000 years, as shown in the graphs below.

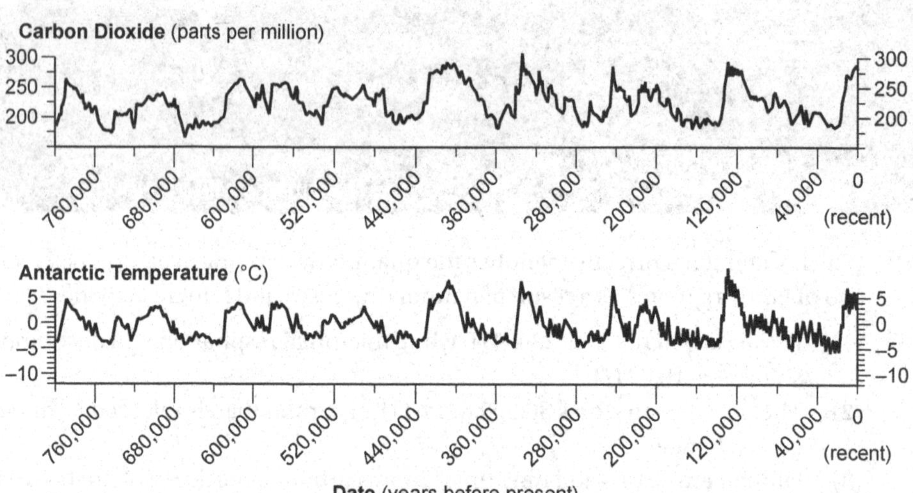

14. Which claim correctly summarizes the data in the graphs that a change to one Earth system caused a change to another Earth system?

(1) Increasing atmospheric CO_2 levels caused the Antarctic surface temperature to decrease over the same time period.
(2) Variations in atmospheric CO_2 levels did not affect the surface temperature in Antarctica over the last 800,000 years.
(3) Decreasing atmospheric CO_2 levels caused Antarctic surface temperatures to also decrease over the same time period.
(4) Atmospheric CO_2 levels stayed the same, causing Antarctic surface temperatures to also stay the same over the last 800,000 years.

14 ______

15. Using the rate of regional climate change in Antarctica for the last 40,000 years, make an evidence-based forecast of how much Antarctic temperatures are predicted to change in the next 40,000 years. Describe a specific associated impact to **one** Earth system as a result of this temperature change. [1]

______________ **°C in next 40,000 years**

Associated impact: ______________________________

____

____

Base your answers to questions 16 through 20 on the information below and on your knowledge of Earth and space sciences. Some questions may require the use of the ***2024 Edition Reference Tables for Earth and Space Sciences****.*

Stars and the Big Bang

Our Sun is a 4.6-billion-year-old yellow dwarf star. It was formed either from matter produced during the Big Bang or from matter released when large stars reached supernova and exploded. This matter, in the form of hydrogen, contracted into a denser gas cloud due to gravity. The temperature at this cloud's core increased, allowing for the fusion of two hydrogen nuclei into one helium nucleus. The mass of this helium nucleus is slightly less than the mass of the hydrogen nuclei. This difference in mass is the source of the star's energy.

The model below shows some information about the life cycles of different stars.

Model of Life Cycles of Different Stars

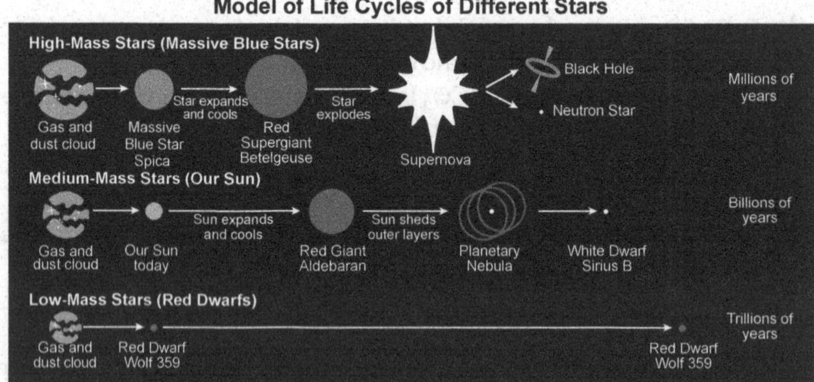

16. Identify the factor that determines the lifespan of the Sun and other stars. [1]

____

The table below shows some information about different types of stars. Solar mass is the mass of the star compared to the Sun.

Star	Solar Mass	Distance from Sun (light years)	Approximate Lifespan (yr)
Spica	10.3	260.9	less than 30 million
Betelgeuse	16.5	548	10 million
Sun	1.0	0	9 billion
Aldebaran	1.16	65	6.4 billion
Sirius B	0.98	8.6	0.23 billion
Wolf 359	0.09	7.86	4.1 trillion

17. A student created a data table containing information on how stars synthesize common elements through nucleosynthesis. Which row of data correctly identifies all the characteristics for that star?

Row	Star Name	Nucleosynthesis	Mass	Lifespan (yr)
(1)	Sun	hydrogen $\xrightarrow{\text{changes directly to}}$ helium	medium mass	4.6 billion
(2)	Sirius B	carbon $\xrightarrow{\text{changes directly to}}$ oxygen	high mass	0.23 billion
(3)	Aldebaran	helium $\xrightarrow{\text{changes directly to}}$ carbon	medium mass	6.4 billion
(4)	Wolf 359	hydrogen $\xrightarrow{\text{changes directly to}}$ carbon	high mass	4.1 trillion

17 ______

Sunspots are areas where the magnetic field is about 2500 times stronger than Earth's magnetic field. Because of this strong magnetic field, the magnetic pressure increases and the Sun's surrounding atmospheric pressure decreases. This lowers the temperature relative to surrounding areas because it inhibits the flow of new super-hot gas (plasma) to the surface.

Sunspots occur in pairs because they have magnetic fields pointing in opposite directions. However, from 1645 to 1715 there was nearly zero sunspot activity. This time period is referred to as the Maunder Minimum. Some scientists also called this time period on Earth "The Little Ice Age."

The graph below shows some information about sunspot frequency.

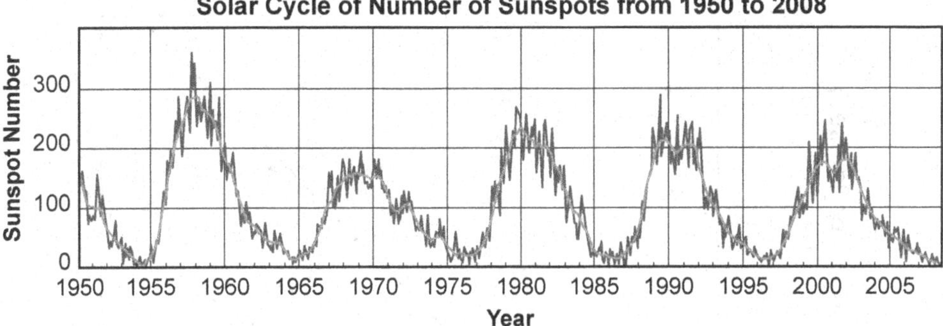

18. Using information from the passage and the graph, place a check mark (✓) in **three** boxes to identify the statements that accurately describe sunspots. [1]

☐ A decrease in the number of sunspots is inferred to decrease Earth's temperatures.

☐ The number of sunspots changes each year, occurring in approximately 11-year cycles.

☐ An increase in solar output is associated with a decrease in the number of sunspots.

☐ The average number of sunspots appearing each year has decreased steadily since 1950.

☐ Sunspots are regions of cooler temperatures on the surface of the Sun.

In the 1920s, Edwin Hubble studied the motion of galaxies. He found a relationship between a galaxy's velocity as measured from Earth (recessional velocity) and the galaxy's distance from Earth. This relationship is known as Hubble's Law. This law has implications for understanding how the universe has changed since the Big Bang.

The graph below shows data on several galaxies' distances from Earth in megaparsecs (Mpc) and their recessional velocities.

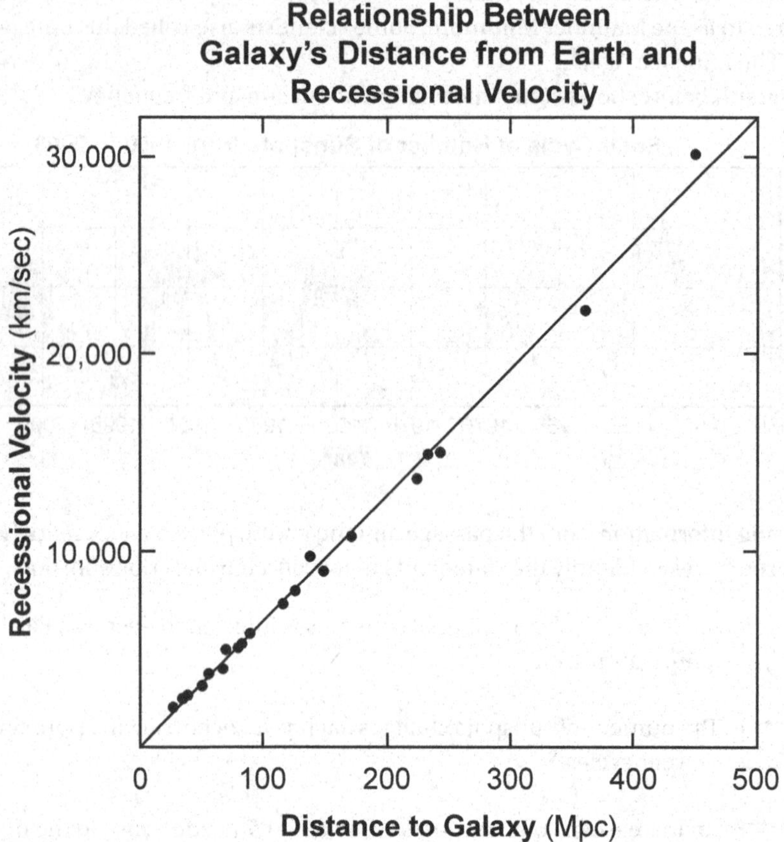

The *Relationship Between Galaxy's Distance from Earth and Recessional Velocity* graph shows that the recessional velocity of a galaxy __A__ as the galaxy's distance from Earth increases. This is evidence for the __B__ of the universe and suggests that the universe initially was __C__ at the time of the Big Bang. As a result, these data suggest that the universe is changing at __D__ rate.

19. Which table below correctly identifies the missing words and phrases labeled *A*, *B*, *C*, and *D* in the passage above?

(1)

A	increases proportionally
B	expansion
C	compacted
D	an accelerated

(3)

A	increases proportionally
B	expansion
C	inflated
D	a decreasing

(2)

A	increases nonproportionally
B	expansion
C	inflated
D	a constant

(4)

A	increases nonproportionally
B	expansion
C	compacted
D	an accelerated

19 ______

Particles were created as a result of the Big Bang. The first particles were subatomic particles like the protons, neutrons, and nuclei of lighter elements such as hydrogen, helium, lithium, and beryllium. The graph below shows some information about these particles, the temperature of the universe at the time they were created, and the amount of time after the Big Bang that they were created.

Relationship Between Relative Amount of Different Particles and Temperature of the Universe After the Big Bang Over Time

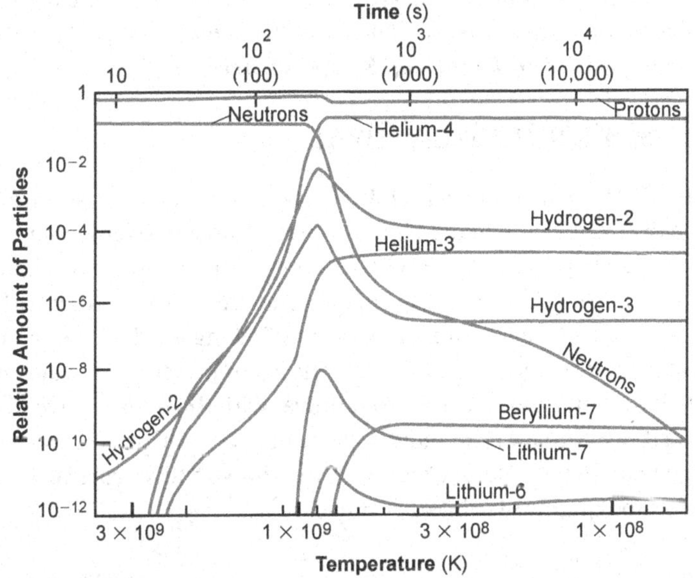

20. Based on information from the *Relationship Between Relative Amount of Different Particles and Temperature of the Universe After the Big Bang Over Time* graph, which table correctly identifies the composition of matter in the universe as evidence for the Big Bang theory?

(1)

From Beginning of Big Bang	Particles Present	Temperature (K)
from 10 to 100 seconds	H and He decreased	increased, then decreased

(2)

From Beginning of Big Bang	Particles Present	Temperature (K)
from 100 to 1,000 seconds	H increased, then decreased while He increased, then remained constant	decreased

(3)

From Beginning of Big Bang	Particles Present	Temperature (K)
from 1000 to 10,000 seconds	protons and neutrons decreased	decreased, then increased

(4)

From Beginning of Big Bang	Particles Present	Temperature (K)
after 10,000 seconds	Be and Li remained constant	remained constant

20 ______

Base your answers to questions 21 through 25 on the information below and on your knowledge of Earth and space sciences. Some questions may require the use of the ***2024 Edition Reference Tables for Earth and Space Sciences.***

New York State's Hudson River

New York State's Hudson River watershed covers almost 1340 square miles and includes three different watersheds: Mohawk River, Hudson River Estuary, and Upper Hudson River watersheds. The Hudson River flows south for almost 325 miles from the Adirondacks to New York City. The 153-mile section from Troy to New York Harbor in New York City is a tidal estuary. Here, fresh water flowing south down the river meets salt water pushing in from the Atlantic Ocean. The leading edge of sea water entering the estuary is called the salt front. The salt front moves with the tides, the weather, and the seasons. When there is heavy rain, more fr esh water flows into the Hudson River. Cities and towns that take their drinking water from the Hudson River carefully track the salt front as it affects the quality of drinking water.

Hudson River Watersheds

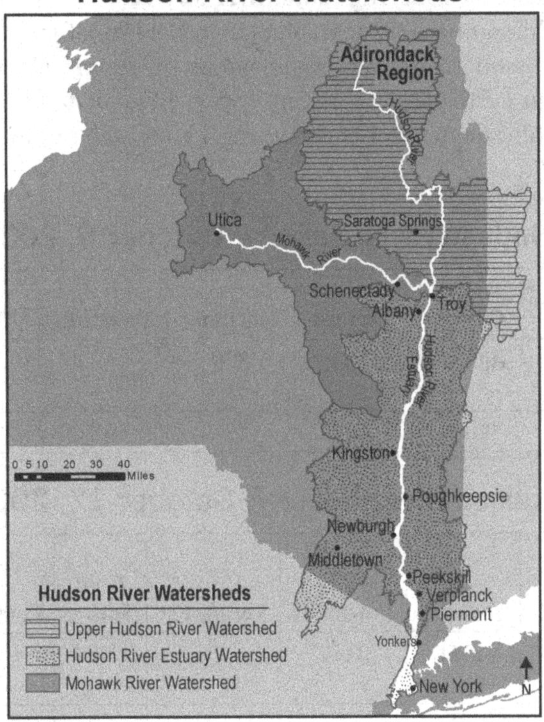

21. Describe how heavy rain events along the Hudson River would affect the location of the salt front. [1]

____

____

____

The location of the salt front is measured in HRM (Hudson River Mile) units. Hudson River Mile 0 is at the southern tip of Manhattan in New York City.

The tables below show some information about the salinity along the Hudson River for several locations north of New York City on two different dates. Salinity is measured in milligrams of chloride per liter of water (mg/L), and the salt front is located where salinity is 100 mg/L.

Hudson River Salinity: October 6, 2004

Location	New York City	Yonkers	Piermont	Bear Mt.	Cold Spring	Ulster
Salinity (mg/L)	1805	1162	300	50	47	34
HRM	7	18	25	46	55	97

Hudson River Salinity: October 12, 2006

Location	New York City	Yonkers	Piermont	Verplanck	Cold Spring	Poughkeepsie	Ulster
Salinity (mg/L)	7362	4041	3177	830	50	30	64
HRM	7	18	25	41	55	76	97

22. A student makes a claim that the location of the salt front is constantly changing due to weather conditions. Which table below supports the student's claim by correctly identifying the two locations between which the salt front was located on October 6, 2004, and October 12, 2006?

(1)

October 6, 2004	Piermont and Bear Mountain
October 12, 2006	Poughkeepsie and Ulster

(2)

October 6, 2004	Piermont and Bear Mountain
October 12, 2006	Verplanck and Cold Spring

(3)

October 6, 2004	Yonkers and Piermont
October 12, 2006	Verplanck and Cold Spring

(4)

October 6, 2004	New York City and Yonkers
October 12, 2006	Yonkers and Piermont

22 ______

An advancing salt front along the Hudson River can affect the quality of drinking water for communities such as Poughkeepsie, which use fresh water from the river as a source of drinking water. For this reason, two different environmental groups carefully monitor the salt front in the river.

Since over 10 million people in New York State rely on the Hudson River for clean drinking water, both environmental groups have developed a plan to evaluate salt front advancement and to prevent the salt from entering the drinking water intakes.

This $400,000 plan has a budget to monitor and evaluate salt front advancement over the next 35 years. This budget has two components summarized below:

- Allocate $250,000 to predict salt front location from 2025 to 2075 using stream and river flow data.
- Allocate $150,000 to develop a proactive action plan to maintain safe drinking water for several water treatment plants.

23. Based on evidence in the plan developed by the environmental groups, which explanation correctly describes how the availability of fresh water will influence communities along the Hudson River?

(1) Cities along the Hudson River will need to spend considerable amounts of money to find alternative drinking water sources as the salinity decreases over the next 50 years.

(2) Cities along the Hudson River will need to spend $400,000 to monitor the salt front over the next 50 years.

(3) Communities that use the Hudson River for drinking water will spend $400,000 to monitor salinity and develop plans to purify water if necessary at water treatment plants.

(4) Communities that use the Hudson River for drinking water will need to spend $150,000 to predict whether the salt front will affect their drinking water.

23 ______

The location of the salt front and its effect on drinking water quality is not the only issue facing residents along the Hudson River.

Between 1947 and 1977, industries that manufactured substances used in fire prevention and oil insulators, called PCBs (polychlorinated biphenyls), were found to be toxic to humans and life in the river. By then, the industry, located north of Albany, had dumped an estimated 1.3 million pounds of PCBs into the river. Once in the river, the chemicals mixed with sediments on the river bottom and along the shorelines. The removal of a dam in the upper Hudson in 1973 further released large amounts of contaminated sediments.

In 1984, the Environmental Protection Agency (EPA) classified a 200-mile stretch of the river as a federal Superfund site that required the removal of the PCBs from the river sediment. A 40-mile stretch north of Albany was termed as a "hot spot" where PCB contaminated sediment was removed by dredging between 2009 and 2015. The model below shows some information on PCBs in the Hudson River.

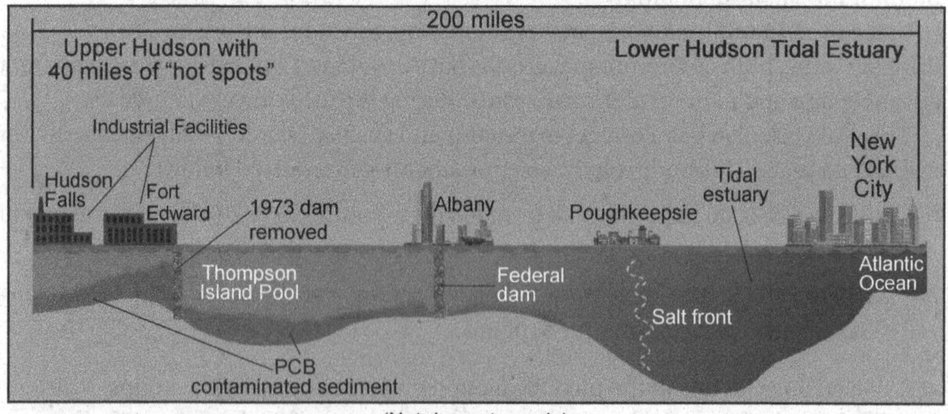

(Not drawn to scale)

Eating contaminated fish is the single greatest human exposure to PCBs. The PCBs in the upper Hudson River have been present for 70 years and have accumulated in fish.

The EPA issued an advisory to not eat fish taken from the upper Hudson River. In 2002, the EPA adopted targets of PCB concentrations in fish to be reached by 2020 and 2031. The graph below shows PCB concentrations of fish taken from the upper Hudson River and these targets.

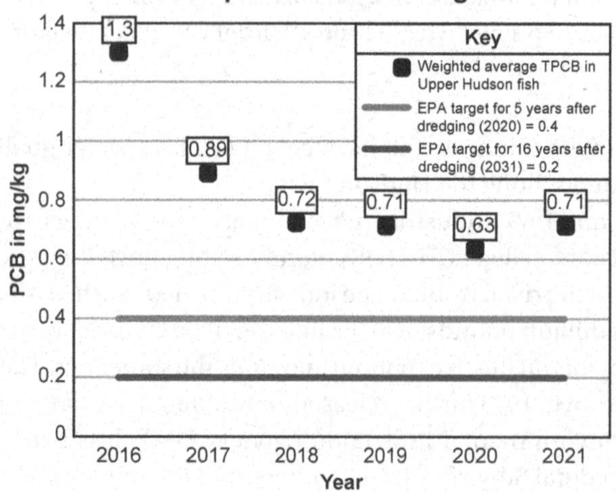

24. Based on the model and graph, which would be the next logical solution that would cause the *greatest* decrease in PCB levels in fish in the Hudson River?

(1) Do nothing and let the river naturally remove the PCBs into the Atlantic Ocean.

(2) Dredge additional areas that still have PCBs in sediment to permanently remove pollutants from the river.

(3) Build another large dam where the dam was removed to capture most PCBs before the contaminated sediment moves downstream.

(4) Breed larger fish that can remove PCBs from the water and sediments.

24 ______

25. Explain how the graph supports the claim that dredging of the Hudson River has only been partially effective at reducing PCB levels in fish, compared to EPA targets. [1]

______

______

______

Base your answers to questions 26 through 30 on the information below and on your knowledge of Earth and space sciences. Some questions may require the use of the ***2024 Edition Reference Tables for Earth and Space Sciences.***

The Origin of Our Solar System

The Sun and eight planets in the Solar System formed at the same time. Evidence of their formation can be found throughout the Solar System. Scientists have used data from planets, meteorites, and Earth to determine how the Solar System formed and its early history. The data table below lists some information for the eight planets.

Solar System Data

Type of Planet	Name	Number of Moons
Terrestrial planets	Mercury	0
	Venus	0
	Earth	1
	Mars	2
Jovian planets	Jupiter	80 (approx.)
	Saturn	83 (approx.)
	Uranus	24
	Neptune	14

26. Which statement best describes the differences between terrestrial planets and Jovian planets in the Solar System as a result of the early history of their formation?

(1) Terrestrial planets have longer periods of revolution and fewer moons than Jovian planets.

(2) Terrestrial planets have greater densities and longer periods of revolution than Jovian planets.

(3) Terrestrial planets have larger diameters and are closer to the Sun than Jovian planets.

(4) Terrestrial planets have smaller diameters and greater densities than Jovian planets.

26 ______

27. Which value would be the most accurate prediction for the period of revolution for Saturn?

(1) 5358 Earth days

(2) 10,759 Earth days

(3) 23,560 Earth days

(4) 28,286 Earth days

27 ______

Meteorites on Earth have been helpful in reconstructing the history of our Solar System and our planet. Samples taken from meteorites have been age dated using absolute dating techniques that measure the amount of uranium-238 compared to the amount of its decay product in samples found on Earth. The graph below shows some information about uranium-238.

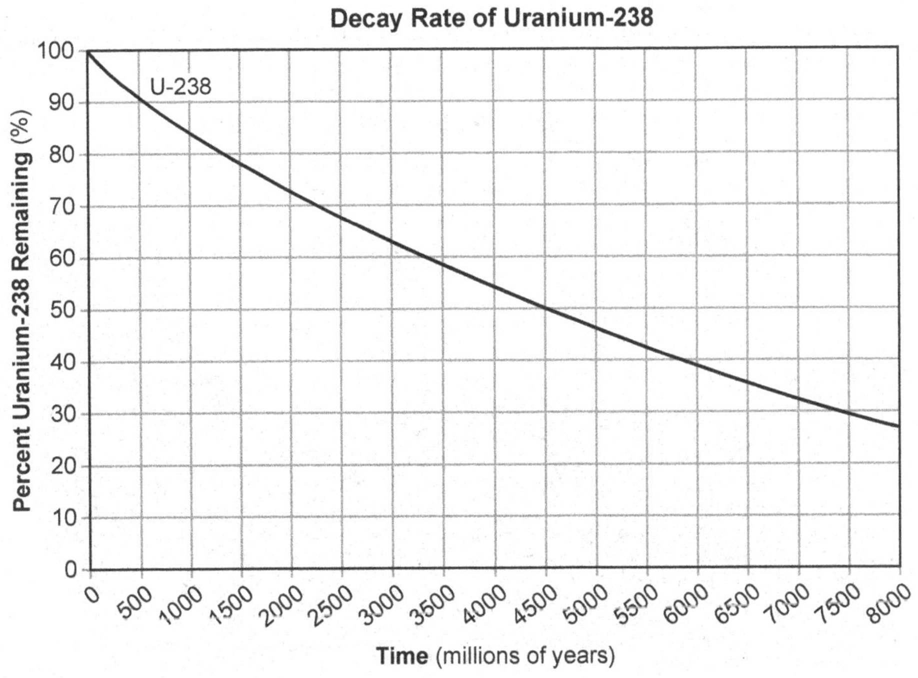

28. A sample of a meteorite was tested and found to contain 50% uranium-238. Use evidence from the graph to make a claim about how the radiometric dating of meteorites can be used to construct an account about when Earth formed. [1]

____

____

____

Throughout Earth's history, meteorites have created impact craters on continental and ocean crust surfaces. However, of the approximately 200 craters confirmed on Earth's surface, only about 20 are located within ocean crust. This is surprising since approximately 70% of Earth's surface is covered in water. It is suggested by scientists that the movement of Earth's plates may play a role in the lack of evidence of impact craters observed on ocean crust.

The approximate rates of movement of oceanic and continental crust are indicated on the model in cm/yr. Two impact craters are labeled.

Model of Plate Tectonic Movement

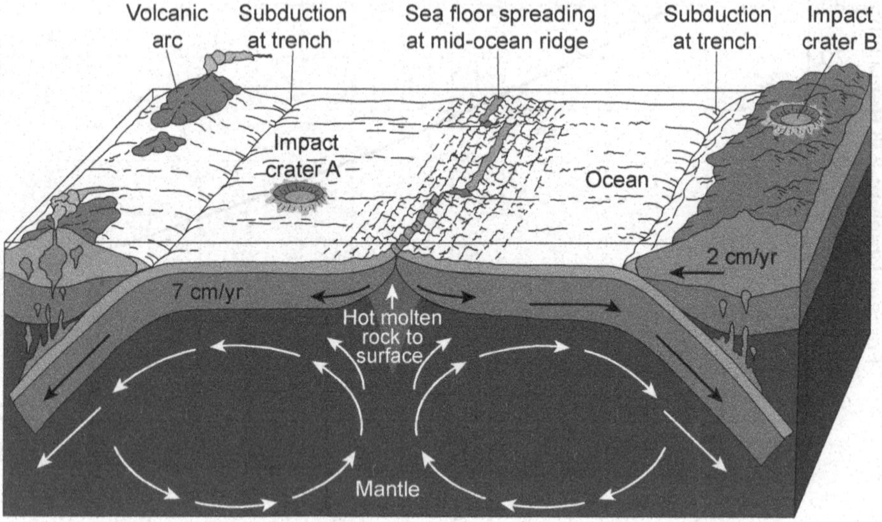

(Not drawn to scale)

29. Which statement most accurately identifies why evidence of impact crater *A* will most likely *not* be preserved for as long as impact crater *B*?

(1) Impact crater *A* is located on faster moving ocean crust and will be destroyed by subduction.

(2) Impact crater *A* is located on a slower moving plate and will be subducted before impact crater *B*.

(3) Impact crater *A* will be destroyed by hot molten rock at the mid-ocean ridge.

(4) Impact crater *A* will be destroyed by eruptions from the nearby volcanic arc.

29 ______

Earth's systems do not operate independently. For instance, changes in the ocean's temperature can influence atmospheric temperatures. A similar coevolutionary process occurred in Earth's early history when oxygen levels were transitioning from Earth's ocean to the atmosphere.

The model below shows some information about Earth's ocean and atmosphere.

Model of Changing Oxygen Levels Through Earth's History

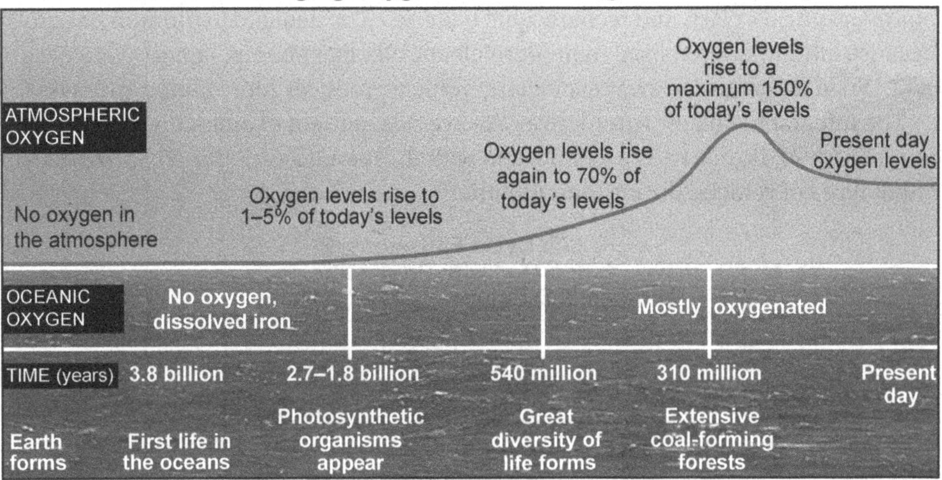

30. Identify the process that caused the oxygen levels in Earth's oceans and atmosphere to change. Then, construct an argument, based on evidence, that describes how oxygen levels changed in Earth's oceans and atmosphere *and* how these changes led to a coevolution of life on two Earth spheres. [1]

Process: ______________________________

Argument with evidence: ______________________________

____

____

Base your answers to questions 31 through 35 on the information below and on your knowledge of Earth and space sciences. Some questions may require the use of the ***2024 Edition Reference Tables for Earth and Space Sciences.***

Mining and Use of Lithium

Lithium is a highly reactive alkali metal that is used in the manufacturing of lubricants, pharmaceuticals, glass, and rechargeable batteries. The demand for lithium has grown because lithium is often used in modern electronics like phones, laptops, and electric cars. When in operation, a battery electric vehicle produces zero tailpipe emissions.

The infographic below summarizes the average amount of emissions, measured in tons of carbon dioxide (tCO_2e) associated with the manufacturing and use of three different types of vehicles during their lifetime.

Life Cycle Emissions Associated with Production and Use of Different Types of Vehicles

	Battery Electric Vehicle	Hybrid Electric Vehicle	Internal Combustion Engine Vehicle
Battery Manufacturing	5 tCO_2e	1	
Vehicle Manufacturing	9 tCO_2e	9 tCO_2e	10 tCO_2e
Fuel/Electricity Production	26 tCO_2e	12 tCO_2e	13 tCO_2e
Tailpipe Emissions		24 tCO_2e	32 tCO_2e
Maintenance	1	2	2
Total	41 tCO_2e	48 tCO_2e	57 tCO_2e

31. Identify the type of vehicle from the infographic that has the ***smallest*** negative impact on the environment. Justify your response by providing evidence from the infographic. [1]

Type of vehicle: ______________________________

Evidence: ______________________________

____

The data table below summarizes originally proposed toll rates for motorists entering New York City by Port Authority bridges and tunnels from New Jersey. Vehicles eligible for the Green Pass discount include plug-in hybrid electric vehicles and battery electric vehicles.

Bridge and Tunnel Tolls for Motorists Entering New York City from New Jersey

Class	Vehicle Type	# of Axles	Toll Off-peak Hours Eastbound toll only	Toll Peak Hours Eastbound toll only
1	**Vehicles with two axles and single rear wheels** (includes two axle recreational vehicles with single rear wheels and no additional axles in tow)	2	$12.75	$14.75
7	**Class 1 or 11 (including class 1 recreational vehicles) with trailer** (minimum three single wheel axles)	3 and Up	$24.25 Additional axles $11.50 each	$26.25 Additional axles $11.50 each
Discount Plans (Enrollment Required)	**Green Pass** Eligible low-emission class 1 vehicles	2	$9.25 Additional axles $11.50 each	$14.75 Additional axles $11.50 each
	Green Pass Eligible low-emission class 7 vehicles	2	$20.75 Additional axles $11.50 each	$26.25 Additional axles $11.50 each

32. Vehicles entering New York City add pollutants to the air in a congested urban area. Describe the economic benefit to motorists who drive class 1 and class 7 vehicles during off-peak hours that qualify for the Green Pass toll rate. Also, discuss how the wants and needs of society are affected by the Port Authority toll plan. [1]

Economic benefit: ______________________________

____

Wants and needs of society: ______________________________

____

____

Lithium is currently sourced from surface mines or underground brine reservoirs. Lithium ore extracted from open pit mines is dried in fossil fuel burning kilns. The underground reservoir brine method uses evaporating ponds and solar energy to collect lithium from brine water having a high concentration of various salts. The table below shows some information related to these two mining methods.

	Mining Methods	
	Mine	Reservoir Brine Method
Emission of Carbon Dioxide (per 1000 kg of lithium)	15,000 kg	5000 kg
Use of Water (per 1000 kg of lithium)	170 m^3	469 m^3
Use of Land (per 1000 kg of lithium)	464 m^2	3124 m^2

33. Which table below correctly summarizes the benefits of extracting lithium by mining or the reservoir brine method?

(1)

Benefits	Mining	Reservoir Brine Method
lower emission of CO_2		✓
less use of water	✓	
less use of land		✓

(3)

Benefits	Mining	Reservoir Brine Method
lower emission of CO_2	✓	
less use of water		✓
less use of land	✓	

(2)

Benefits	Mining	Reservoir Brine Method
lower emission of CO_2	✓	
less use of water		✓
less use of land		✓

(4)

Benefits	Mining	Reservoir Brine Method
lower emission of CO_2		✓
less use of water	✓	
less use of land	✓	

33 ______

Lithium production has impacts on land use in the Atacama region of Chile. The photograph shows solar evaporation ponds at a brine extraction and processing facility.

The graph below shows some information about lithium.

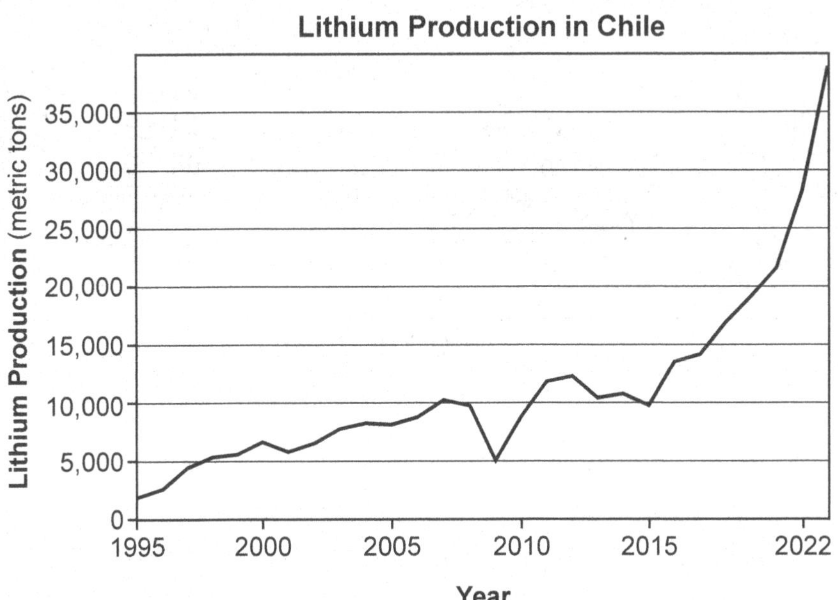

34. Which statement correctly identifies the changes to the lithosphere that were needed for reservoir brine lithium extraction in Chile between 1995 and 2022?

(1) The amount of land surface needed for evaporation ponds increased at a consistent rate from 2005 to 2010.

(2) The amount of land surface needed for evaporation ponds was lowest in 1995 and greatest in 2015.

(3) The amount of land surface needed for evaporation ponds increased between 2015 and 2022.

(4) The amount of land surface needed for evaporation ponds only decreased from 2010 to 2015.

34 ______

The Andean flamingo, a bird native to wetlands in the Atacama region of Chile and other locations where lithium deposits occur, is listed as a vulnerable species. The model below shows some information about the reservoir brine method. The graph shows projected changes from 2020 to 2060 in Atacama wetland areas due to lithium mining.

Model of Reservoir Brine Method

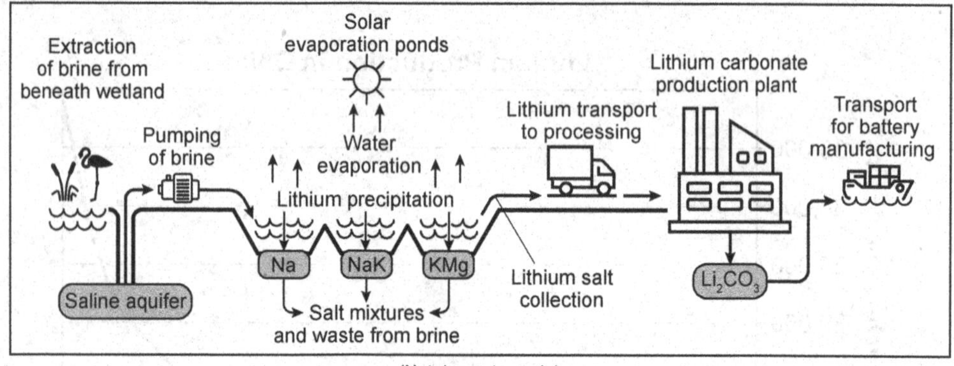

(Not drawn to scale)

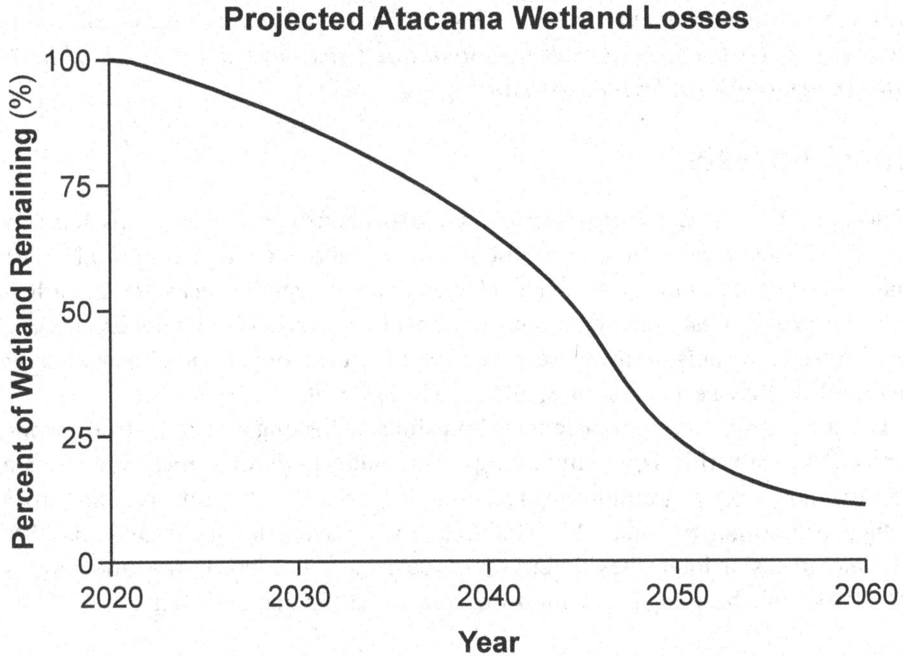

35. Which statement correctly identifies how the processing and management of saline brine for lithium production in the Atacama region of Chile negatively impacts the Andean flamingo habitat?

(1) Pumping of brine from solar evaporation ponds adds brine to wetlands.

(2) Pumping of brine from saline aquifers reduces water that would be available for wetlands.

(3) Evaporation of water from saline aquifers reduces water that would be available for wetlands.

(4) Evaporation of water from solar evaporation ponds adds water to wetlands.

35 ______

Base your answers to questions 36 through 40 on the information below and on your knowledge of Earth and Space Sciences. Some questions may require the use of the ***2024 Edition Reference Tables for Earth and Space Sciences.***

Ghost Forests

"Ghost forest" is a term used to describe a group of dead trees or tree stumps that remain standing. Ghost forests often form due to environmental change or natural disasters that impact coastal forests. As sea level changes, the invading seawater can advance and overtake the fresh water that many tree species in coastal forests need to exist. This salty water slowly poisons living trees, leaving dead and dying timber. Ghost forests can be found in almost every coastal state in the United States.

Coastal forests serve as efficient carbon sinks, collecting and storing atmospheric carbon. They are critical for maintaining water quality as they naturally filter, cool, and slow the movement of groundwater and streams. Coastal forests protect against erosion, buffer storm surges, provide wildlife habitats, and ensure water quality and quantity. As saltwater intrusion intensifies, the supply of coastal wood needed by the timber industry will also shrink, harming the economy of rural areas that depend on it.

Coastal Ghost Forest

The graphs below show some information about global temperature, CO_2, and sea level. An anomaly is a change from an expected value. An anomaly value of 0 represents no change from the historical average.

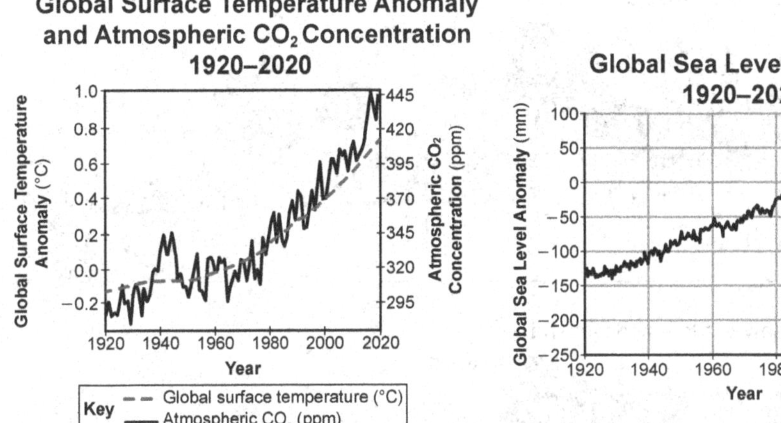

36. In addition to increased carbon dioxide emissions, which row in the table below correctly describes the factors and their effect on coastal flooding in areas where ghost forest land cover has increased from 1920 to 2020?

Row	Total Change in Temperature Anomaly	Total Global Sea Level Anomaly	Coastal Flooding
(1)	−0.9°C	70 mm	decreased
(2)	−0.9°C	195 mm	decreased
(3)	0.9°C	195 mm	increased
(4)	0.9°C	70 mm	increased

36 ______

The infographic below shows the predicted development of a ghost forest in a coastal area near a residence from the year 2020 to 2100. Brackish water is a mix of fresh and salty water.

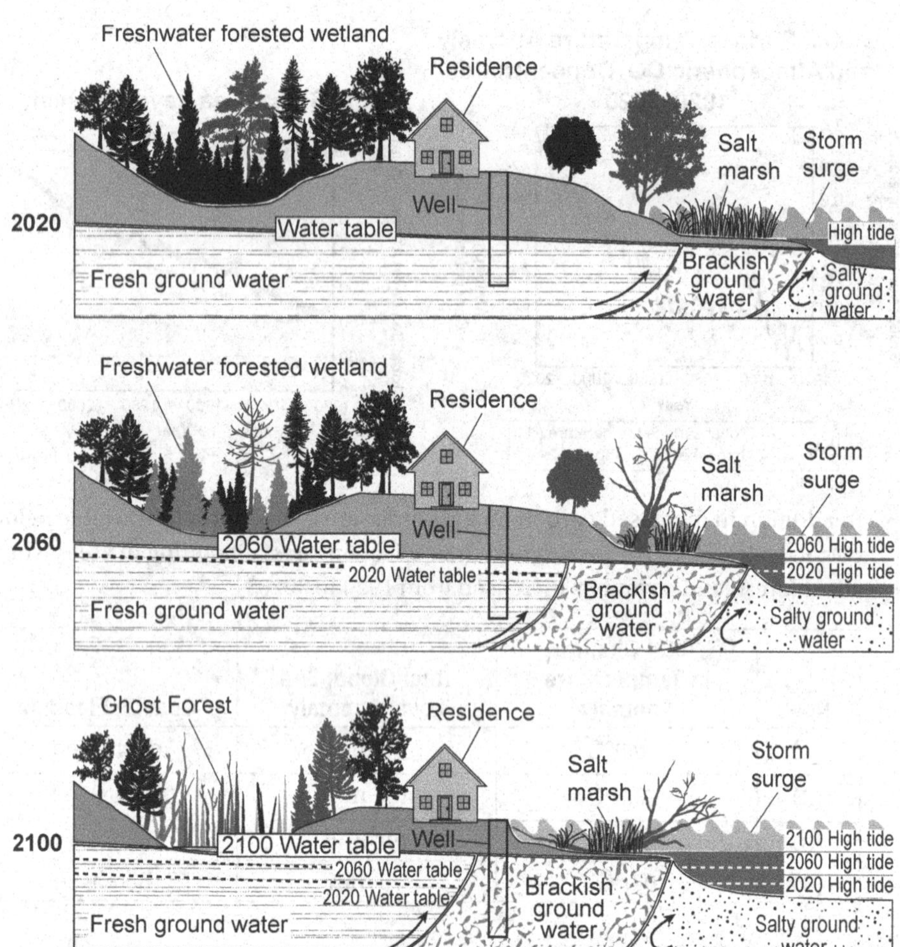

37. Based on the infographic, identify **one** natural resource that will be affected by the rise in sea level, and construct an explanation about how the change in the availability of this resource will affect human activity. [1]

Natural resource: ______________________________

Effect on human activity: ______________________________

____

38. A student makes the claim that the amount of living forest acreage will decrease as the amount of ghost forest acreage increases in the coastal area near this residence. Which associated impact to an Earth system will most likely occur?

(1) The increase in wetland area will decrease local humidity.

(2) The loss of ground vegetation will cause an increase in water runoff and soil erosion.

(3) As high tide water levels increase, the local climate will cool due to more solar energy being reflected.

(4) New species that live on dead and decaying wood in ghost forests will increase the biodiversity of the ecosystem.

38 ______

The model below shows the effects of subsidence in the creation of ghost forests as a result of an earthquake that occurred on January 26, 1700, along the west coast of the United States.

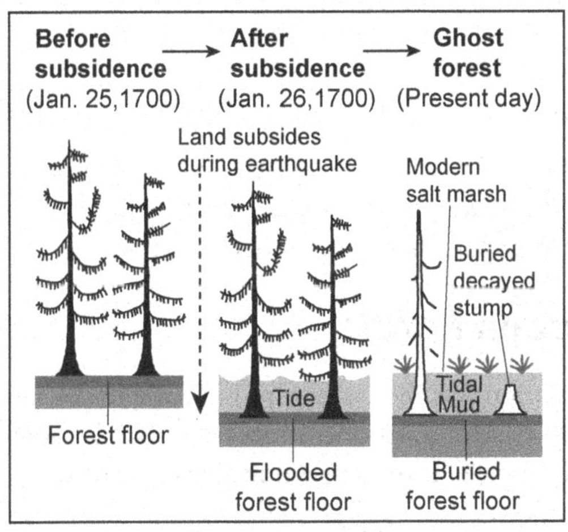

(Not drawn to scale)

39. Use the model to compare the temporal scale of land subsidence associated with the earthquake to the temporal scale of land subsidence associated with the development of a ghost forest. [1]

Temporal scale of land subsidence associated with an earthquake: ______________

____

Temporal scale of land subsidence associated with the development of a ghost forest: ____

____

____

Six proposed solutions for reducing the expansion of ghost forests are below.

Proposed Solutions

(1) Fortify seawalls with concrete
(2) Use sand and soil to create dunes with grasses along the coast
(3) Limit development along coastal wetlands
(4) Install sand fencing to reduce loss of sand from wind erosion
(5) Preserve and restore the biodiversity of vegetation in tidal areas
(6) Install tall rock walls and other artificial breakwaters

40. Which three proposed solutions reduce the expansion of ghost forests and would most likely be accomplished with minimal environmental impact and maximum aesthetic value to local residents?

(1) 1, 3, 4 (3) 2, 5, 6
(2) 1, 4, 6 (4) 2, 3, 5

40 ______

Base your answers to questions 41 through 45 on the information below and on your knowledge of Earth and space sciences. Some questions may require the use of the ***2024 Edition Reference Tables for Earth and Space Sciences.***

Human Impact on Earth

Global human population has increased dramatically in the last 50 years, which has increasingly stressed Earth's natural resources. This has created challenges in the management of those natural resources while also creating opportunities to find solutions to those challenges.

The EN-ROADS simulator is an online tool that allows users to manipulate variables in order to see their effects on climate change. A student used this simulator to estimate the global temperature change by the year 2100 if no changes to current conditions were made and if changes from current conditions were made.

The graphs show the results of the simulations. *Simulation 1* shows the possible outcome of greenhouse gas net emissions and the associated atmospheric global temperature change under current conditions. *Simulation 2* shows the possible outcome of greenhouse gas net emissions and the associated atmospheric global temperature change due to a different set of computer input conditions.

Simulation 1: Greenhouse Gas Net Emissions– No Change from Current Human Practices

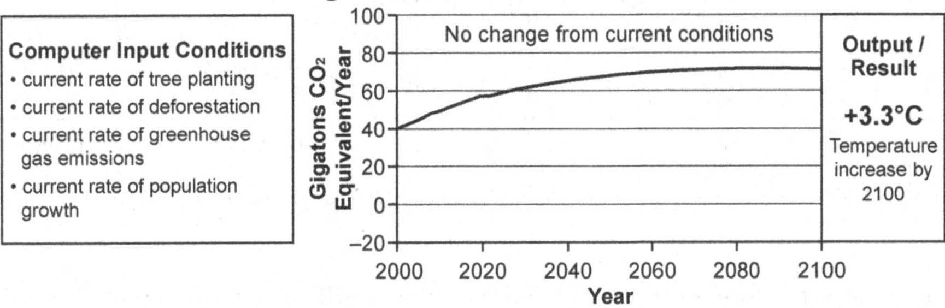

Simulation 2: Greenhouse Gas Net Emissions– Changed Human Practices

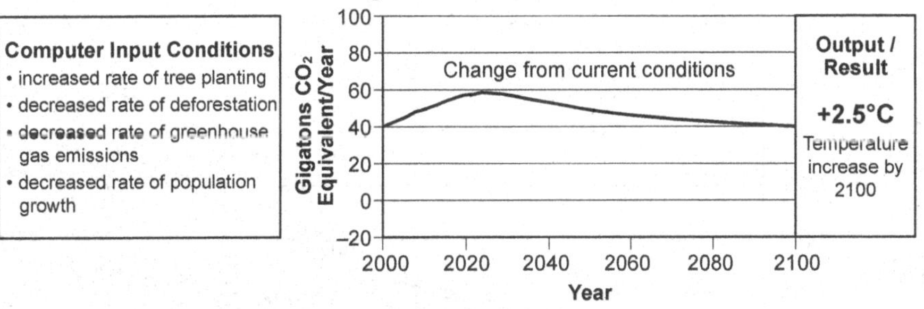

41. Based on the information from *Simulation 1* and *Simulation 2*, which row in the table below correctly matches both a method of managing natural resources and the resulting effect on human population, greenhouse gases in the atmosphere, and biodiversity beginning in 2020?

Row	Method of Managing Natural Resources	Effect on Human Population	Amount of Greenhouse Gases in Atmosphere	Effect on Biodiversity
(1)	replanting forests	positive	decrease	positive
(2)	replanting forests	positive	no change	negative
(3)	decreasing deforestation	positive	no change	positive
(4)	decreasing deforestation	positive	decrease	negative

41 ______

The graphs below show some information about different types of fossil fuels.

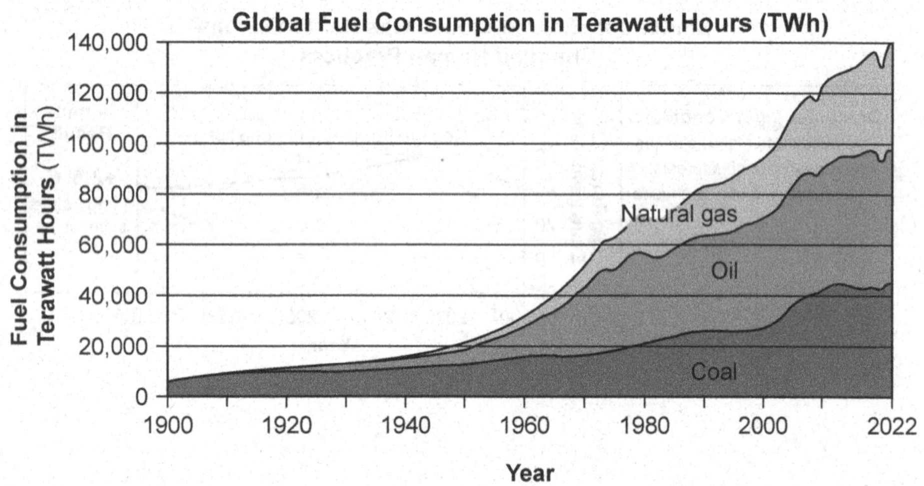

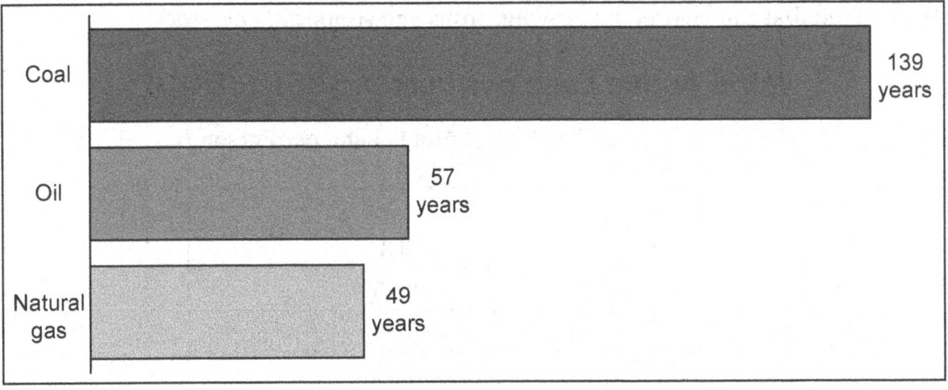

42. Based on the information in both graphs, which statement correctly identifies the relationship between the management of fossil fuels since 1950 and the future sustainability of these resources?

(1) The yearly consumption of fossil fuels has increased by more than six times and the production of some types of these resources is not sustainable beyond the year 2150.

(2) The yearly consumption of fossil fuels has increased by more than ten times and the production of some types of these resources is sustainable beyond the year 2170.

(3) The yearly consumption of fossil fuels has doubled and the production of some types of these resources is sustainable for the next 30 years.

(4) The yearly consumption of fossil fuels has tripled and the production of some types of these resources is not sustainable for the next 30 years.

42 ______

The table below shows some information about arable land from 1961 to 2016. Arable land is land that can be used for growing crops and supporting livestock.

World Arable Land per Person 1961 to 2016

Year	**Hectares of Arable Land per Person** (1 hectare = 2.47 acres)
1961	0.36
1972	0.3
1983	0.26
1994	0.25
2005	0.21
2016	0.19

43. Based on the graphs and table, which statement correctly provides evidence to support the claim that increased consumption of fossil fuels caused a change in the amount of world arable land per person?

(1) Atmospheric temperature increased, which increased soil evaporation and increased the amount of arable land.

(2) Atmospheric temperature increased, which decreased soil evaporation and decreased the amount of arable land.

(3) Atmospheric temperature increased, which decreased soil evaporation and increased the amount of arable land.

(4) Atmospheric temperature increased, which increased soil evaporation and decreased the amount of arable land.

43 ______

The Cheetah Conservation Fund (CCF) is a nonprofit organization that is working to ensure the survival of the cheetah and its habitat in Namibia, Africa. The central part of this African country was once a mixed woodland savannah. However, overgrazing by livestock (cattle raised for meat), hunting of elephants and rhinos, and removal of mature trees has changed the landscape to be dominated by thornbushes. This restricts the ability of animals in this habitat to hunt. CCF harvests these thornbushes to make Bushblok, a miniature log which can be used as a biomass fuel source.

44. Describe how the manufacturing of Bushblok logs addresses a problem in Namibia. Explain how these logs benefit the cheetah population and reduce the impact of human activity on cheetah habitat in Namibia. [1]

Problem: ____

____

____

Benefit: ____

____

____

45. Increasing the use of biomass is one of many actions that the CCF is taking to reduce the rate of climate change. Which additional solution, if *increased,* could be used along with biomass as another renewable energy source?

(1) Drilling for oil
(2) Mining of natural gas
(3) Use of solar panels
(4) Burning woodland trees

45 ______

Base your answers to questions 46 through 50 on the information below and on your knowledge of Earth and space sciences. Some questions may require the use of the ***2024 Edition Reference Tables for Earth and Space Sciences.***

Tectonics of the Hawaiian Islands

The Hawaiian Islands are situated near the center of the Pacific Plate and are volcanic peaks representing a history of volcanic eruptions. The island of Hawaii is over a hotspot, where a magma source in the mantle pushes upward to Earth's surface creating active volcanoes. The island of Hawaii itself is still being formed by ongoing volcanic activity at Mauna Loa and Kilauea volcanoes, both of which are currently situated over the hotspot. Loihi, an undersea volcano, also sits above the hotspot and will likely become the next Hawaiian island. The other islands in the chain are extinct volcanoes.

The maps show some information about the Hawaiian islands. Varying ages of bedrock on many of the islands are indicated in millions of years. The model shows three stages of island formation over a mantle hotspot.

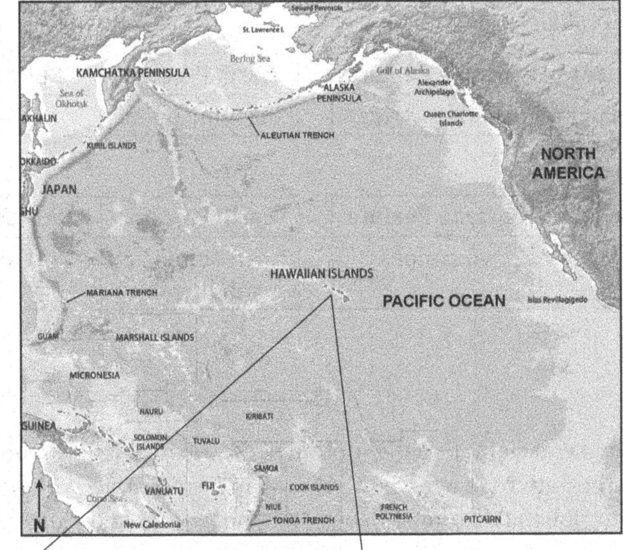

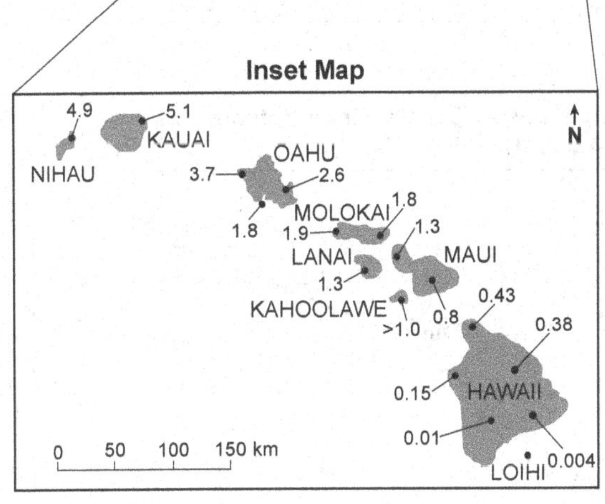

Model

Ocean surface

Pacific Plate

A

Stage 1

Magma source

Active volcano

Ocean surface

A B

Stage 2

Extinct volcanoes

A B C

Stage 3

(Not drawn to scale)

46. Which table correctly pairs the Earth process with the surface feature that created the Hawaiian Islands chain?

Earth Process	Surface Feature
sinking magma	mid-ocean ridge

(1)

Earth Process	Surface Feature
rising magma	volcanoes

(2)

Earth Process	Surface Feature
tectonic uplift	volcanoes

(3)

Earth Process	Surface Feature
tectonic subduction	oceanic trench

(4)

46 ______

47. A student makes a claim that the motion of the Pacific Plate can be used to determine the pattern of ages of the Hawaiian Islands. Use evidence from the maps to describe the relationship between the pattern in the ages of the islands and the compass direction that the Pacific Plate is moving. [1]

Pattern in ages: ______________________________

Compass direction of movement: ______________________________

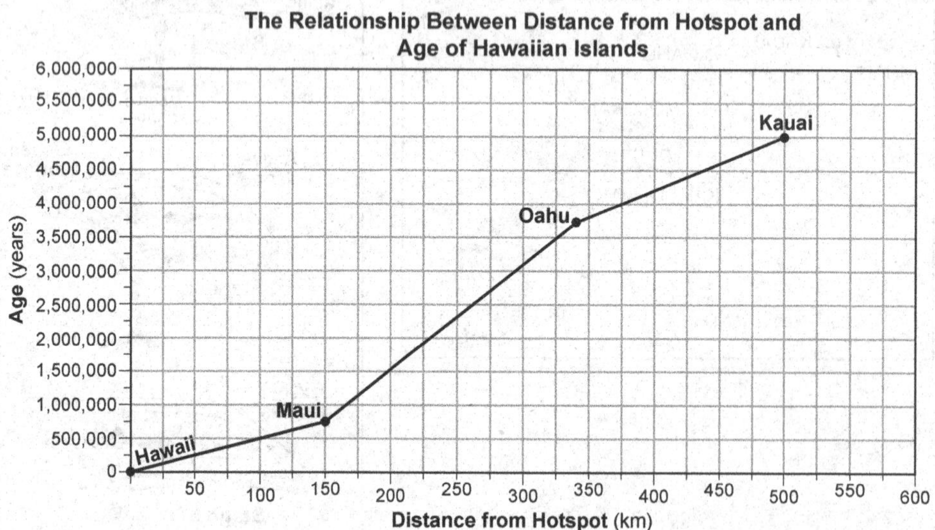

Two samples of ocean crust were collected between some of the Hawaiian Islands. Data for samples *X* and *Y* are shown below.

Sample	Approximate Age (y)	Approximate Distance (km)
X	500,000	90
Y	2,750,000	275

48. Between which islands were the samples collected?

(1) *X*—between Hawaii and Maui
Y—between Oahu and Kauai

(2) *X*—between Oahu and Kauai
Y—between Hawaii and Maui

(3) *X*—between Hawaii and Maui
Y—between Maui and Oahu

(4) *X*—between Maui and Oahu
Y—between Hawaii and Maui

48 ______

The models below show the geologic history of the formation of present-day Maui.

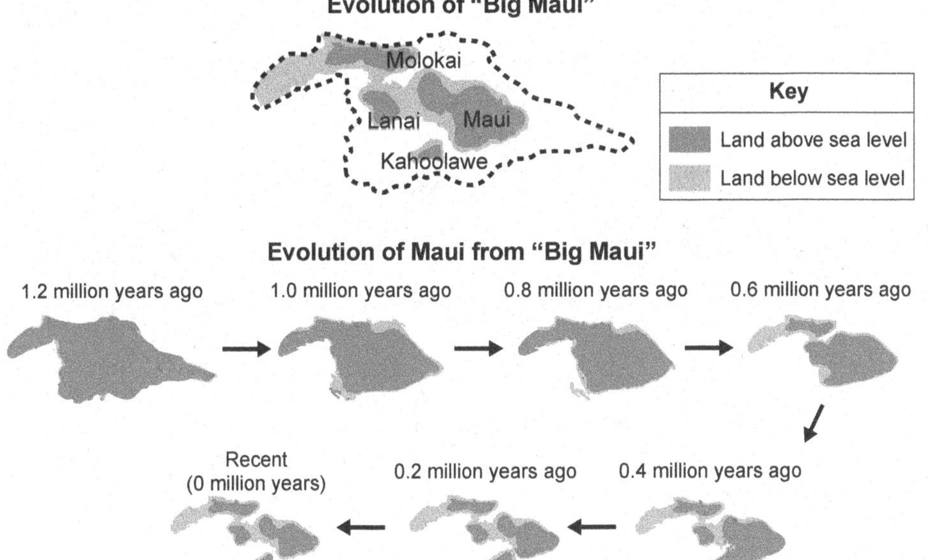

49. Based on the models, which table correctly identifies the spatial change, the temporal change, and one possible surface process that contributed to the evolution of Big Maui? [1]

(1)

Spatial Change	Surface Feature	Surface Process
1 large island to 4 separated islands	1.2 billion years	sea level rise

(2)

Spatial Change	Surface Feature	Surface Process
4 separated islands to 1 large island	4.2 million years	deposition along shorelines

(3)

Spatial Change	Surface Feature	Surface Process
1 large island to 4 separated islands	1.2 million years	sinking of landmass

(4)

Spatial Change	Surface Feature	Surface Process
4 separated islands to 1 large island	4.2 billion years	sea level drop

49 ______

The photographs below show some shoreline locations on the island of Maui. Human development along the shoreline has impacted coastal regions. Scientists designed solutions to reduce this impact.

Photograph 1: Shoreline Location 1

Photograph 2: Shoreline Location 2

Photograph 3: Shoreline Location 3
Along Coastal Highway

50. Identify the geoscience problem being addressed in the photographs. Describe how the solutions shown in photograph 2 *or* photograph 3 reduce the impact of human development on natural geoscience processes that occur in coastal regions. [1]

Geoscience problem: ______________________________

How solutions reduce impact: ______________________________

____

____

Answer Explanations June 2025

1. **Sample Response:**

   - The Sun should be placed on the main sequence, in spectral class G, near 5800 K and absolute magnitude +5.
   - As it transitions to a red giant:
     - Change in relative temperature: Decreases
     - Change in relative absolute magnitude: Decreases

Explanation

When the Sun becomes a red giant, its surface cools as it expands, so its temperature decreases. Simultaneously, it becomes brighter, causing the absolute magnitude value to decrease since lower numbers mean brighter objects.

2. **(4)** The correct sequence for energy flow from the core to the outer solar layers is core → radiative zone → convective zone → photosphere → chromosphere. This pathway reflects the natural movement of energy outward from where it is created (core) to where it is released as visible light (photosphere) and then to the outer atmospheric layers (chromosphere).

Wrong Choices Explained

(1) The chromosphere is above the photosphere and does not directly conduct energy outward from the core.

(2) This option omits the critical convective zone, which is necessary for heat transfer before reaching the photosphere.

(3) This option incorrectly places the photosphere before the corona and the transition region in the wrong order.

3. **Correct Response:**

   - ✓ Mercury travels faster in its orbit when it is closer to the Sun.
   - ✓ Venus's orbit is less elliptical than Mercury's orbit.

Explanation

Mercury's orbit has a higher eccentricity (0.206), meaning it moves faster when closer to the Sun due to Kepler's second law. Venus's orbit is nearly circular (0.007), making it less elliptical than Mercury's.

4. (2) According to Kepler's third law, planets farther from the Sun orbit more slowly. A planet between Mercury and Venus would have an average speed that is less than Mercury's but greater than Venus's.

Wrong Choices Explained

(1) Mercury, being closer to the Sun, moves faster than both the hypothetical planet and Venus, making this choice incorrect.

(3) The hypothetical planet is inside Venus's orbit and therefore cannot orbit slower than Venus.

(4) Earth orbits beyond Venus, so its orbital speed is slower than that of the hypothetical planet, contradicting the choice.

5. Correct Response:

- Choice *A*: inside
- Choice *B*: closer to
- Choice *C*: Sun

Explanation

Venus orbits inside Earth's orbit, meaning it's sometimes closer to Earth and sometimes on the far side of the Sun. These changing positions allow observers on Earth to see phases of Venus, just like they see phases of the Moon.

6. (3) When Earth's obliquity (tilt) is at a minimum (22.1°), the northern hemisphere receives more evenly distributed solar energy throughout the year, which reduces seasonal extremes. During winter, this tilt results in more energy reaching polar regions, decreasing ice formation.

Wrong Choices Explained

(1) Less energy would cause more, not less, ice formation.

(2) Less energy leading to more ice contradicts the expected energy outcome of decreased obliquity.

(4) More energy results in less ice, not more.

7. **(2)** Glacial melting at point *X* introduces fresh water into the ocean, decreasing the salinity and density of the water. This disrupts the thermohaline circulation, leading to a weakening of ocean currents like the Gulf Stream.

Wrong Choices Explained

(1) Adding fresh water does not make currents deeper.

(3) Less salty water is less dense, not denser.

(4) Fresh water does not warm the current directly; it affects density.

8. **(1)** Location *Y* is affected by warm ocean currents that transport heat from the tropics. These warm currents raise air temperature and increase evaporation, which results in warmer air and more precipitation.

Wrong Choices Explained

(2) Warmer air usually increases precipitation, not decreases it.

(3) and (4) Location *Y* experiences warm currents, not cool ones.

9. **(4)** As glaciers melt, dark surfaces like rock or water are exposed, which absorb more solar energy. This increases warming and leads to more glacial melting, continuing a positive feedback loop.

Wrong Choices Explained

(1) The process starts with decreased glacial ice, not increased glacial ice as shown in this choice.

(2) This choice shows the incorrect sequence of effects. The dark surface area should increase after ice melts, not before.

(3) Glacial melting leads to decreased ice, not decreased melting.

10. **Correct Response:**

- Projected CO_2 equivalent in 2100: ~800 ppm
- Projected global surface temperature change: ~2.5°C

Explanation

Using the provided graph for Scenario B, the intersection at the year 2100 shows that greenhouse gas concentrations are near 800 ppm, with a corresponding global temperature rise of about 2.5°C.

11. **Sample Response:** Trees use the process of photosynthesis to produce food. During this process, trees take in carbon dioxide from the atmosphere and use sunlight to convert it into sugars. As a result, atmospheric carbon dioxide levels decrease because trees store the carbon in their tissues.

12. **(4)** According to the model, 92 gigatons (92 GT) of carbon flow naturally between the atmosphere and ocean in both directions—into and out of the ocean. This equilibrium supports the choice that the carbon exchange is balanced at that rate.

Wrong Choices Explained

(1) Fossil carbon does not release 2300 GT naturally; this is a stored value.

(2) The deep ocean stores carbon, but this is not relevant to the 92 GT exchange.

(3) Human emissions do not exceed natural respiration by nine times.

13. **(1)** Replanting trees in burned forest areas is a human activity designed to remove carbon dioxide from the atmosphere. Plants absorb CO_2 during photosynthesis, so replanting helps reduce greenhouse gas levels.

Wrong Choices Explained

(2) Burning fossil fuels increases, not decreases, atmospheric CO_2.

(3) Migration does not reduce carbon dioxide. Migration is an adaptation, not a mitigation strategy.

(4) Using dead trees as fuel would increase emissions.

14. **(3)** The graph shows that as atmospheric CO_2 decreases, Antarctic surface temperatures also decrease. This is direct evidence that a change in the carbon cycle causes a change in temperature.

Wrong Choices Explained

(1) Increasing CO_2 increases temperature, not decreases it.

(2) The graph shows a clear correlation; this claim contradicts the data.

(4) CO_2 levels and temperatures both vary; they do not stay the same.

15. Correct Response:

- °C in next 40,000 years: 5°C
- Associated impact: Melting of polar ice sheets, which leads to rising sea levels and coastal flooding in low-lying areas.

Explanation

From the ice-core data, we observe an ~5°C increase in Antarctic temperature over the past 40,000 years. If the current trend continues, a similar rise is expected, leading to further melting of ice and impacting the hydrosphere.

16. Correct Response: The factor that determines the lifespan of the Sun and other stars is the mass of the star.

Explanation

The greater the mass of a star, the faster it burns through its nuclear fuel. Low-mass stars burn fuel slowly and live longer, while high-mass stars burn nuclear fuel quickly and have shorter lifespans.

17. **(3)** Aldebaran is a medium-mass star currently in the red giant phase. In this stage, it fuses helium into carbon through nucleosynthesis. With a lifespan of approximately 6.4 billion years, this classification is consistent with the expected life cycle of a medium-mass star evolving off the main sequence.

Wrong Choices Explained

(1) The Sun is a medium-mass star, but it is still on the main sequence and currently fuses hydrogen into helium—not directly into carbon. The table incorrectly claims it fuses hydrogen directly into carbon.

(2) Sirius B is a white dwarf and no longer undergoes any form of fusion. It does not perform carbon-to-oxygen nucleosynthesis as an active process, and the lifespan of 0.23 billion years is far too short for its complete stellar evolution.

(4) Wolf 359 is a low-mass red dwarf—not a high-mass star as stated. It fuses hydrogen into helium at a very slow rate and never becomes hot enough to fuse hydrogen directly into carbon. The lifespan given (4.1 trillion years) is consistent for red dwarfs, but the nucleosynthesis and mass classification are incorrect.

18. Correct Response:

- ✓ A decrease in the number of sunspots is inferred to decrease Earth's temperatures.
- ✓ The number of sunspots changes each year, occurring in approximately 11-year cycles.
- ✓ Sunspots are regions of cooler temperatures on the surface of the Sun.

Explanation

Sunspots affect solar output. Fewer sunspots correlate with cooler periods, such as the Maunder Minimum. Sunspots also follow a cycle and are cooler than surrounding areas due to magnetic activity.

19. **(1)** As a galaxy's distance increases, its recessional velocity increases proportionally, supporting Hubble's Law. This is evidence of the expansion of the universe, which implies that the universe was once compacted and is now accelerating in expansion.

Wrong Choices Explained

(2) Hubble's Law shows a proportional, not a nonproportional, increase.

(3) "Inflated" is not accurate; "compacted" better describes the early universe.

(4) "Nonproportionally" contradicts the data trend in the graph.

20. **(2)** From 100 to 1000 seconds after the Big Bang, hydrogen (H) increased and then decreased, while helium (He) increased and then remained constant. During this same period, the temperature steadily decreased, allowing the formation of stable atomic nuclei such as helium-3 and helium-4.

Wrong Choices Explained

(1) From 10 to 100 seconds, the graph shows that hydrogen and helium were increasing, not decreasing. Also, the temperature did not increase and then decrease—it only decreased continuously after the Big Bang.

(3) From 1000 to 10,000 seconds, protons and neutrons had already stabilized, and their relative amounts were not significantly decreasing. The temperature also did not increase during this phase; it continued to decrease, contradicting the claim.

(4) After 10,000 seconds, the relative abundances of beryllium-7 and lithium-7 were extremely low and did not remain strictly constant. More importantly, this

period is beyond the major nucleosynthesis window and not representative of the main composition of matter used as evidence for the Big Bang.

21. **Sample Response:** Heavy rain events would increase the amount of fresh water flowing into the Hudson River. This would push the salt front farther south toward the Atlantic Ocean because the added fresh water would displace the incoming salt water.

22. **(2)** On October 6, 2004, the salt front was between Piermont and Bear Mountain. By October 12, 2006, it had moved northward, between Verplanck and Cold Spring, indicating a shift due to changing flow of fresh water.

Wrong Choices Explained

(1) The salt front was beyond Poughkeepsie by 2006.

(3) Yonkers is too far south.

(4) The salt front didn't stay near New York City and Yonkers in either year.

23. **(3)** The environmental plan includes monitoring the salt front ($250,000) and creating an action plan for water purification ($150,000). These efforts help protect communities dependent on Hudson River water.

Wrong Choices Explained

(1) Salinity is expected to increase, not decrease.

(2) Monitoring alone doesn't address purification.

(4) $150,000 is for planning, not prediction.

24. **(2)** Dredging additional contaminated areas would reduce PCB levels by permanently removing pollutants, which accumulate in fish tissues.

Wrong Choices Explained

(1) "Do nothing" delays remediation.

(3) Dam building doesn't remove PCBs already in sediments.

(4) Breeding fish doesn't remove toxins from the environment.

25. **Sample Response:** The graph shows that from 2016 to 2021, the average PCB levels in fish remained above the EPA's target level of 0.4 mg/kg, indicating that dredging has not fully removed PCBs from the Hudson River. Therefore, the cleanup has been only partially effective.

26. **(4)** Terrestrial planets (like Earth and Mars) have smaller diameters and greater densities compared with gas giants (Jovian planets). This difference arose from early Solar System formation; heavier materials stayed closer to the Sun.

Wrong Choices Explained

(1) Terrestrial planets have shorter periods of revolution.

(2) Terrestrial planets have greater densities but not longer periods of revolution.

(3) Jovian planets are larger than terrestrial planets.

27. **(2)** Saturn is a planet farther from the Sun than Earth. According to Kepler's third law, Saturn's period of revolution would be longer. The most accurate estimate of its revolution period is 10,759 Earth days.

Wrong Choices Explained

(1) 5358 Earth days is far too short for a planet as far out as Saturn.

(3) 23,560 Earth days is closer to the orbit of Uranus or Neptune.

(4) 28,286 Earth days greatly exceeds Saturn's actual orbital period.

28. **Sample Response:** The graph shows that 50% of uranium-238 remains after about 4.5 billion years, which is one half-life of U-238. If a meteorite contains 50% of its original uranium-238, it means the meteorite is approximately 4.5 billion years old. Scientists use this information to determine that the Earth and Solar System also formed about 4.5 billion years ago, since meteorites are leftover material from the early Solar System.

29. **(1)** Impact crater *A* lies on fast-moving oceanic crust that is actively being subducted. Over time, subduction will destroy evidence of the impact, making it less likely to be preserved than impact crater *B*.

Wrong Choices Explained

(2) Impact crater *B* is on slower-moving crust and is not being subducted as quickly.

(3) Impact crater *A* is not near the mid-ocean ridge, so it will not be destroyed by ridge activity.

(4) Although volcanic activity can affect surface features, subduction is the direct cause of the disappearance of impact crater *A*.

30. Process: Photosynthesis

Argument with evidence: Oxygen levels in Earth's oceans and atmosphere began to rise due to the activity of photosynthetic organisms, such as cyanobacteria. These organisms produced oxygen as a by-product of photosynthesis. As ocean oxygen increased, atmospheric oxygen followed, allowing for the coevolution of life in both the biosphere (animals on land) and atmosphere (oxygen-breathing organisms).

31. Type of vehicle: Battery electric vehicle

Evidence: The battery electric vehicle produces the least total CO_2 emissions over its lifetime according to the infographic. Although its manufacturing emissions are slightly higher due to battery production, the lack of tailpipe emissions and lower fuel-related emissions make the battery electric vehicle more environmentally friendly overall.

32. Economic benefit: Motorists who drive class 1 and class 7 low-emission vehicles during off-peak hours pay lower tolls (e.g., \$9.25 instead of \$12.75), thereby saving money.

Wants and needs of society: The toll discount encourages drivers to switch to low-emission vehicles, reducing urban air pollution and promoting environmental sustainability. This meets society's need for cleaner air and less traffic congestion.

33. **(4)** The reservoir brine method results in lower carbon dioxide emissions (5000 kg vs. 15,000 kg), which is a major environmental benefit. The mining method uses less water and less land, requiring only 170 m^3 of water and 464 m^2 of land compared to 469 m^3 and 3124 m^2 for the brine method. Therefore, each method has distinct environmental advantages, and choice 4 correctly reflects this by assigning the correct benefit to each method.

Wrong Choices Explained

(1) This table incorrectly shows that mining uses less water and land, which is true, but it fails to assign lower CO_2 emissions exclusively to the brine method, which is its main environmental strength.

(2) This table suggests all three benefits apply to both methods, which is inaccurate. The mining method does not emit less CO_2, and the brine method does not use less water or land.

(3) This table incorrectly gives the brine method credit for using less land, when in fact it uses significantly more than mining (3124 m^2 vs. 464 m^2), based on the data provided.

34. **(3)** Lithium production in Chile has increased rapidly between 2015 and 2022, as shown by the production graph. This means the land surface needed for evaporation ponds also increased during this time.

Wrong Choices Explained

(1) The growth wasn't consistent between 2005 to 2010.

(2) The amount of land surface needed didn't peak in 2015 and was surpassed in 2022.

(4) Land use increased, not decreased, between 2010 to 2015.

35. **(2)** The brine is pumped from underground saline aquifers, reducing the amount of water that supports wetland ecosystems. This directly harms habitats like those of the Andean flamingo.

Wrong Choices Explained

(1) Brine is removed, not added, to wetlands.

(3) Evaporation from aquifers isn't the issue. The problem is the extraction.

(4) Water isn't added to wetlands during evaporation; it is lost.

36. **(3)** Between 1920 and 2020, the graphs show a 0.9°C temperature anomaly and 195 mm rise in sea level. This results in increased coastal flooding, which contributes to the expansion of ghost forests.

Wrong Choices Explained

(1) The change was 0.9°C, not –0.9°C.

(2) Coastal flooding has increased, not decreased.

(4) A 70 mm sea level rise underestimates the actual anomaly.

37. Natural resource: Fresh groundwater

Effect on human activity: As the sea level rises, saltwater infiltrates groundwater supplies, making the groundwater unsafe for drinking or household use. People living near the coast may have to rely on expensive water treatment or relocate inland, disrupting communities.

38. **(2)** The loss of living forest leads to loss of ground vegetation, which reduces the soil's ability to absorb water. This results in increased water runoff and soil erosion in ghost forest zones.

Wrong Choices Explained

(1) Wetland area increases do not decrease humidity; vegetation loss reduces it.

(3) An increase in both high tide level and open water absorb more solar energy, thereby warming the local climate.

(4) New species may colonize, but the net loss in biodiversity outweighs the gains.

39. Temporal scale of land subsidence associated with an earthquake: Sudden, occurring in one day (January 26, 1700)

Temporal scale of land subsidence associated with the development of a ghost forest: Gradual, occurring over decades due to sea level rise and saltwater intrusion

40. **(4)** Solutions 2, 3, and 5 (dune building with grasses, limiting development, and restoring tidal biodiversity) are most environmentally friendly and offer aesthetic and ecological benefits.

Wrong Choices Explained

(1) Seawalls and sand fencing are intrusive and visually disruptive.

(2) Seawalls and sand fencing are intrusive and visually disruptive. Concrete and rock walls have major environmental impacts.

(3) Rock walls and artificial breakwaters are the least aesthetic choice and harm coastal processes.

41. **(1)** Replanting forests is a strategy that provides a positive effect on population (more jobs, improved air), decreases atmospheric greenhouse gases, and improves biodiversity by restoring habitats.

Wrong Choices Explained

(2) Replanting forests does change greenhouse gas levels. This option wrongly states "no change."

(3) Decreasing deforestation alone isn't enough to increase biodiversity without replanting.

(4) Deforestation reduction helps biodiversity—not hurts it—thereby making this answer invalid.

42. **(1)** Fossil fuel consumption has increased more than sixfold since 1950. Some reserves, like oil and gas, are projected to be depleted well before 2150, making them unsustainable in the long term.

Wrong Choices Explained

(2) The chart shows a significant increase, but not "more than ten times," and production is not sustainable past 2170.

(3) Consumption increased much more than "doubled."

(4) Consumption tripling is an underestimate, and oil/gas may not last 30 years at this rate.

43. **(4)** As fossil fuel use increased, global temperatures rose, causing increased evaporation from soil, which decreased arable land per person as shown in the declining land chart.

Wrong Choices Explained

(1) Increased evaporation decreases arable land, not increases it.

(2) Higher temperatures do not decrease evaporation; they increase it.

(3) Less evaporation would increase available land, but that didn't happen.

44. Problem: Overgrazing and deforestation in Namibia have caused an overgrowth of thornbushes, limiting hunting grounds and degrading cheetah habitats.

Benefit: Manufacturing Bushblok logs from thornbushes removes these invasive plants, restoring savannah ecosystems for cheetahs and providing a renewable energy source that reduces reliance on wood from mature trees.

45. **(3)** Solar panels are a renewable energy source that produce no emissions and can be used alongside biomass to further reduce fossil fuel dependency.

Wrong Choices Explained

(1) Drilling for oil is nonrenewable and increases carbon emissions.

(2) Mining natural gas releases greenhouse gases.

(4) Burning woodland trees for energy causes deforestation and increases emissions.

46. **(2)** Rising magma from a stationary hotspot beneath the Pacific Plate causes volcanoes to form at the surface as the plate moves over it. This Earth process creates a chain of volcanic islands like the Hawaiian Islands, where newer islands (such as Hawaii and Loihi) are located directly above the hotspot and older islands (like Kauai and Niihau) are extinct and farther away.

Wrong Choices Explained

(1) Sinking magma does not form mid-ocean ridges. In fact, mid-ocean ridges are formed by rising magma at divergent plate boundaries, not at hotspots, and the Hawaiian Islands are not located on a ridge.

(3) Tectonic uplift is responsible for mountain formation, not volcanic island chains. While volcanic islands may rise above sea level due to lava accumulation, uplift is not the primary process here.

(4) Tectonic subduction creates deep ocean trenches and volcanic arcs, typically at convergent plate boundaries, not at intraplate hotspots like the one that created the Hawaiian Islands.

47. Pattern in ages: The farthest islands from the hotspot (e.g., Kauai) are the oldest, while the closest (e.g., Hawaii) are the youngest.

Compass direction of movement: The Pacific Plate is moving in a northwest direction, carrying islands away from the stationary hotspot over time.

48. **(3)** Based on the island age-distance progression, sample *X* (500,000 years, 90 km) lies between Hawaii and Maui while sample *Y* (2,750,000 years, 275 km) lies between Maui and Oahu.

Wrong Choices Explained

(1) Sample *Y* is too old to be between Oahu and Kauai.

(2) Sample *X* is not that far north.

(4) Both sample *X* and sample *Y* are swapped or mislocated in this answer.

49. **(3)** One large island to 4 separated islands is the spatial change observed in the diagram. Over a period of 1.2 million years, the island area once known as "Big Maui" split into the islands of Maui, Molokai, Lanai, and Kahoolawe. This change was caused by the sinking of landmass, which submerged low-lying land below sea level, fragmenting the once-continuous land into separate islands.

Wrong Choices Explained

(1) While the spatial change is correctly identified, the time frame is incorrect—1.2 billion years is far too long. The correct time frame based on the model is 1.2 million years. Also, the primary process was subsidence (sinking) of landmass, not sea level rise.

(2) This reverses the spatial change, incorrectly stating the islands went from 4 to 1. The model clearly shows the opposite trend over time. The time frame (4.2 million years) is also inaccurate, and deposition is not the primary process shown.

(4) Like choice 2, this choice misstates both the direction of spatial change and the timing, and wrongly suggests sea level drop, which would expose more land, not submerge it. The model supports sinking land, not falling sea levels.

50. Geoscience problem: Coastal erosion and shoreline loss due to rising sea levels and wave action

How solutions reduce impact: In photograph 3, the seawall protects the shoreline by deflecting wave energy away from roads and infrastructure, slowing erosion caused by human development near the coast.

Regents Practice Exam August 2025

Directions: **For each multiple-choice question, record in the space provided the number of the choice that best completes the statement or answers the question. For questions requiring a written response, write your answers in the spaces provided.**

Base your answers to questions 1 through 5 on the information below and on your knowledge of Earth and Space Sciences. Some questions may require the use of the ***2024 Edition Reference Tables for Earth and Space Sciences.*** Available at *https://www.nysed.gov/sites/default/files/programs/state-assessment/earth-science-reference-tables-english-2011.pdf*

Energy in the Sun

The Sun's energy influences the environment of all celestial objects in our solar system. Different forms of the hydrogen and helium atoms contained in the Sun's core, deuterium (^{2}H) and the helium atom (^{3}He), are under very high temperatures and pressures. These atoms combine to form helium (^{4}He), while releasing tremendous amounts of energy.

The model below shows some information about the Sun.

Layers of the Sun Model

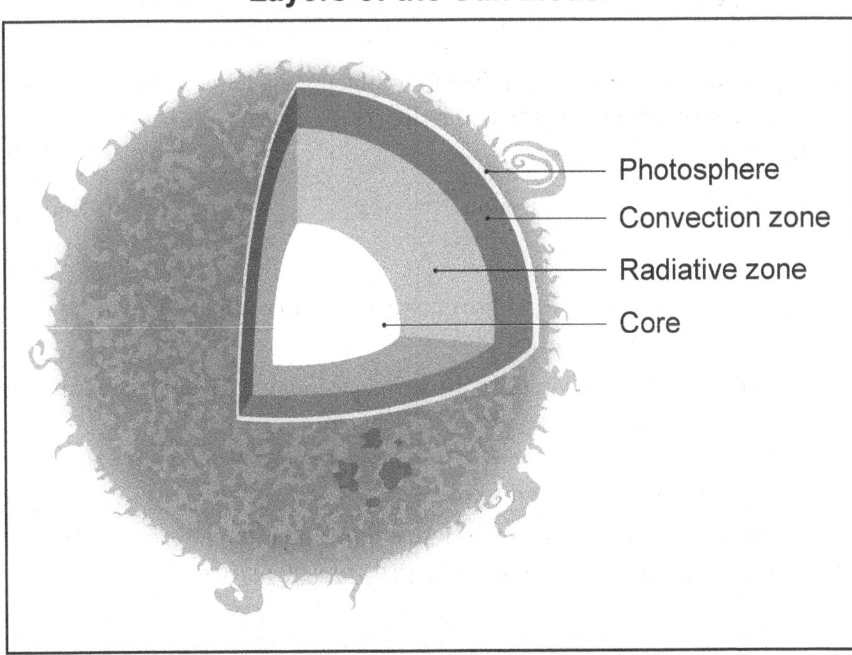

(Not drawn to scale)

Sun's Layer	Information About Layer	Temperature
Photosphere	Observable layer—gives off electromagnetic energy	6700°F–11,000°F or 4000 K–6500 K
Convection zone	• Convection causes hot material to rise to surface and cool • Creates sunspots and solar flares	11,000°F–2 million °F
Radiative zone	Serves as a passage for radiation energy from core to surface	7 million °F
Core	Nuclear reactions occur	27 million °F or 15 million K

Four components of nuclear fusion, labeled *A*, *B*, *C*, and *D*, including models of the nuclei of deuterium, helium-4, and helium-3, are shown below.

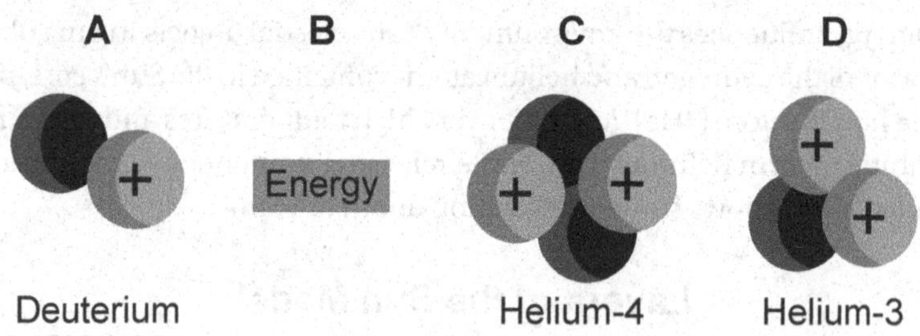

1. Complete the model of nuclear fusion by placing the letter of the component into the correct position in the equation. [1]

$$\_\_\_\_\_\_\_\_ + \_\_\_\_\_\_\_\_ \xrightarrow[\text{Fusion}]{\text{Nuclear}} \_\_\_\_\_\_\_\_ + \_\_\_\_\_\_\_\_ + \text{Neutron}$$

2. Which claim best describes where energy is produced in the Sun and how the energy is released into space?

   (1) Energy is produced in the core, travels through different layers, and is then released at the photosphere.

   (2) Energy is produced in the photosphere, travels through different layers, and is then released at the core.

   (3) Energy is produced in the radiative zone and is directly released into space.

   (4) Energy is produced in the core and is directly released into space.

2 ______

Scientists have identified an 11-year solar cycle. During this cycle there is an increase and a decrease in the number of dark areas (sunspots) on the Sun's surface. Periods in the cycle with the highest numbers are called maxima, while periods in the cycle with the lowest numbers are called minima.

The graphs below show the output of light energy from the entire disk of the Sun, measured at Earth (total solar irradiance) in watts per meter squared (W/m^2), and some information about sunspots.

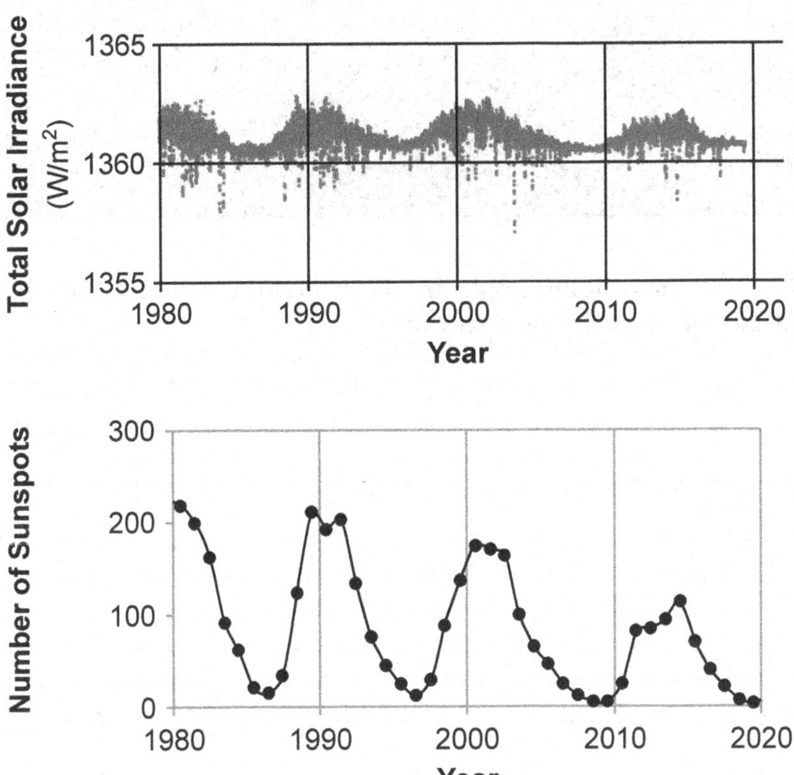

The model below shows photographs of the Sun with active areas (lighter regions) from 2008 to 2017. These active areas indicate high-frequency radiation emitted from the Sun. The line on the model represents the number of sunspots during the same time period.

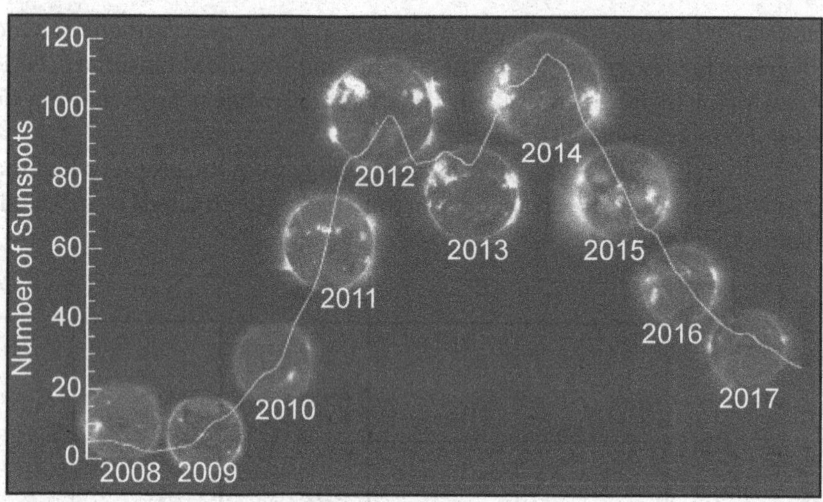

3. Use evidence from the **two** graphs *and* the model to describe how the Sun's irradiance varies with sunspot activity. [1]

___

___

___

4. Which graph best illustrates the projected pattern of the number of sunspots from 2020 to 2040, based on evidence from the graphs and model?

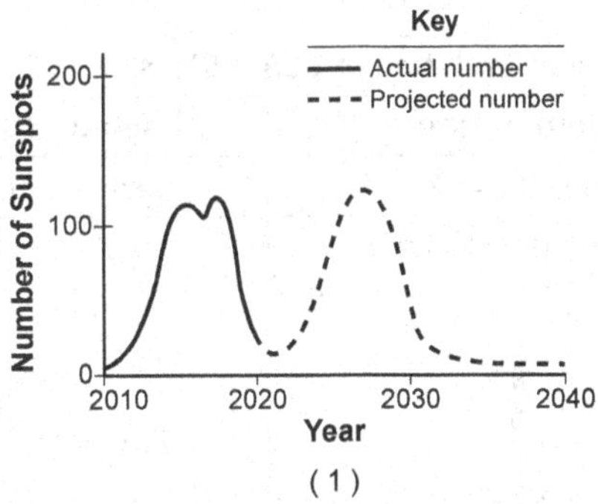

(1)

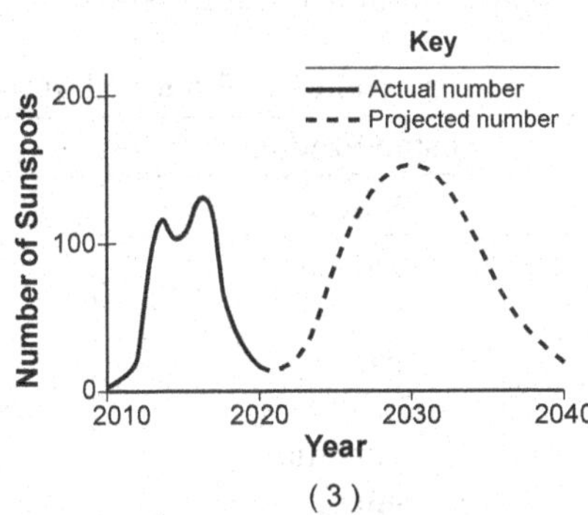

(3)

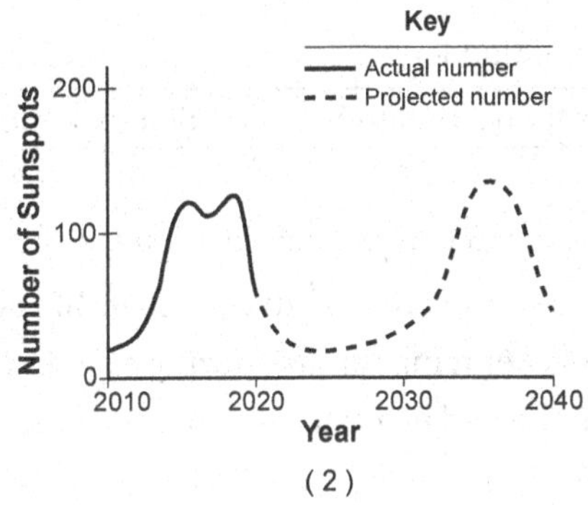

(2)

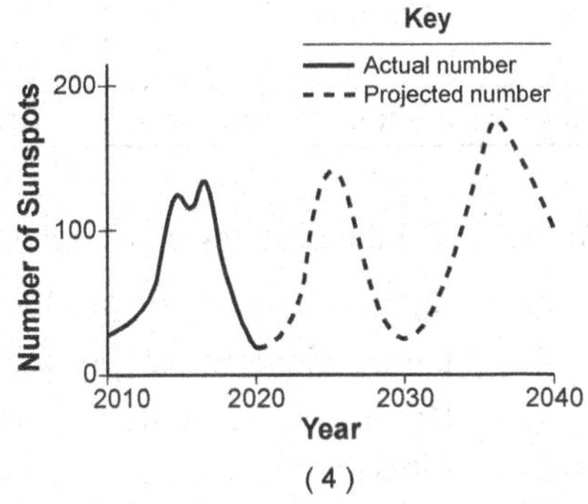

(4)

4 ______

As a star ages, the hydrogen in its core gets used up. The core begins to undergo a gravitational collapse and becomes more dense. The chart below shows some information about elements in some stars.

Elements Produced During Stellar Evolution in Some Stars

Fusion Product	Temperature	Stage of Star
Helium	15 million K	Early stage
Carbon Oxygen Neon Sodium Magnesium Silicon Sulfur Calcium Iron	200 million K ↓ 3 billion K	Late stage
All elements heavier than iron	Greater than 3 billion K	Star explodes

5. Which claim best describes the process of nucleosynthesis in a star?

(1) Lighter elements are produced in a star's core near the end of its life cycle.

(2) Lighter elements are produced at lower temperatures during a supernova.

(3) Heavier elements are produced at the end of a star's life span as temperatures increase.

(4) Heavier elements are produced at lower temperatures and lighter elements are produced at higher temperatures.

5 ______

Base your answers to questions 6 through 10 on the information below and on your knowledge of Earth and Space Sciences. Some questions may require the use of the ***2024 Edition Reference Tables for Earth and Space Sciences.***

The Power of Water

Water in its many forms is unique to our planet. It is required for life as we know it, and it can be used for a wide variety of beneficial purposes such as recreation and generating electricity. However, water can also be destructive and cause problems for humanity.

Concrete is an important building material that can be affected by water. The photograph below shows how structures are built with concrete and rebar (metal rods used to provide strength).

Concrete Pour With Rebar Structure

An experiment was conducted on samples of concrete, with and without rebar, to see how freeze-thaw cycles affected the deformation of the concrete. The graph shows some information about the freeze-thaw experiment.

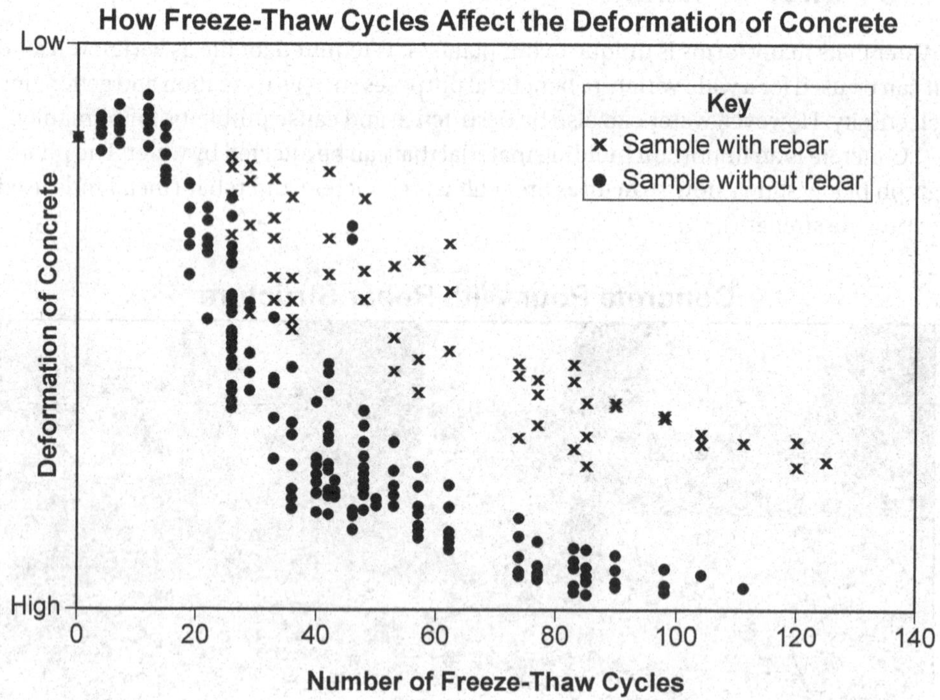

6. Describe a property of water that *increases* the deformation of concrete due to repeated freeze-thaw cycles. Also, describe the relative amount of deformation to concrete samples when metal rebar is added compared to samples without rebar. [1]

Property of water: ______________________________

Relative amount of deformation: ______________________________

____

Calthemite straws are deposits found on the underside of concrete structures such as parking garages. Calthemites are not considered stalactites, as they do not form naturally in caves and cavern systems. The series of photographs below shows the growth of a calthemite straw measured in millimeters (mm).

Growth of Calthemite Straw

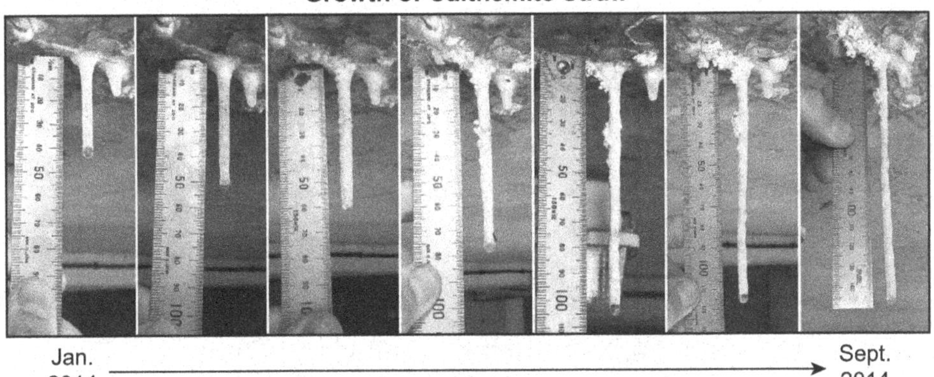

7. Which claim correctly summarizes the effect of water on the concrete in the parking structure producing the calthemite straws?

(1) Surface water evaporates minerals and transports them a short distance. The minerals then condense, forming calthemite straws.

(2) Surfacc watcr mclts minerals and transports them a short distance. The minerals then freeze, forming calthemite straws.

(3) Surface water dissolves minerals and transports them a short distance. The minerals then precipitate, forming calthemite straws.

(4) Surface water absorbs minerals and transports them a short distance. The minerals then expand, forming calthemite straws.

7 ______

The power of water is utilized at Niagara Falls. Two large power generation facilities have been in operation there for decades. Water from the Niagara River is diverted through pipes and tunnels upstream from the falls. This water is then run through turbines to generate electricity before it is released back to the river below the falls.

The map below shows some information about the Niagara Falls region.

Niagara Falls Power Generation

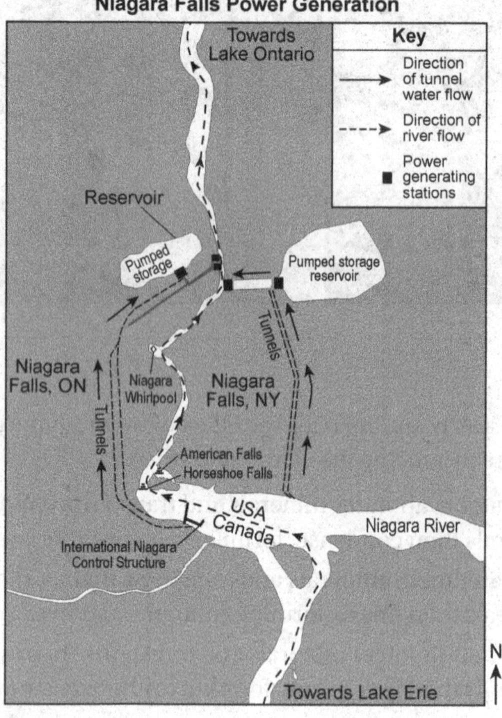

8. How does the production of electricity by hydroelectric power plants at Niagara Falls, rather than by burning fossil fuels, affect Earth's atmosphere and change the climate?

(1) Hydroelectric power plants use a renewable resource that increases atmospheric oxygen levels, which negatively impacts global climate.

(2) Hydroelectric power plants use a renewable resource that decreases atmospheric carbon dioxide levels, which positively impacts global climate.

(3) Hydroelectric power plants use a renewable resource that increases water vapor in the atmosphere, which causes the local climate to be more humid.

(4) Hydroelectric power plants use a renewable resource that decreases water vapor in the atmosphere, which causes the local climate to be more arid.

8 ______

The map below shows some information about the changing position of the edge of the falls over time.

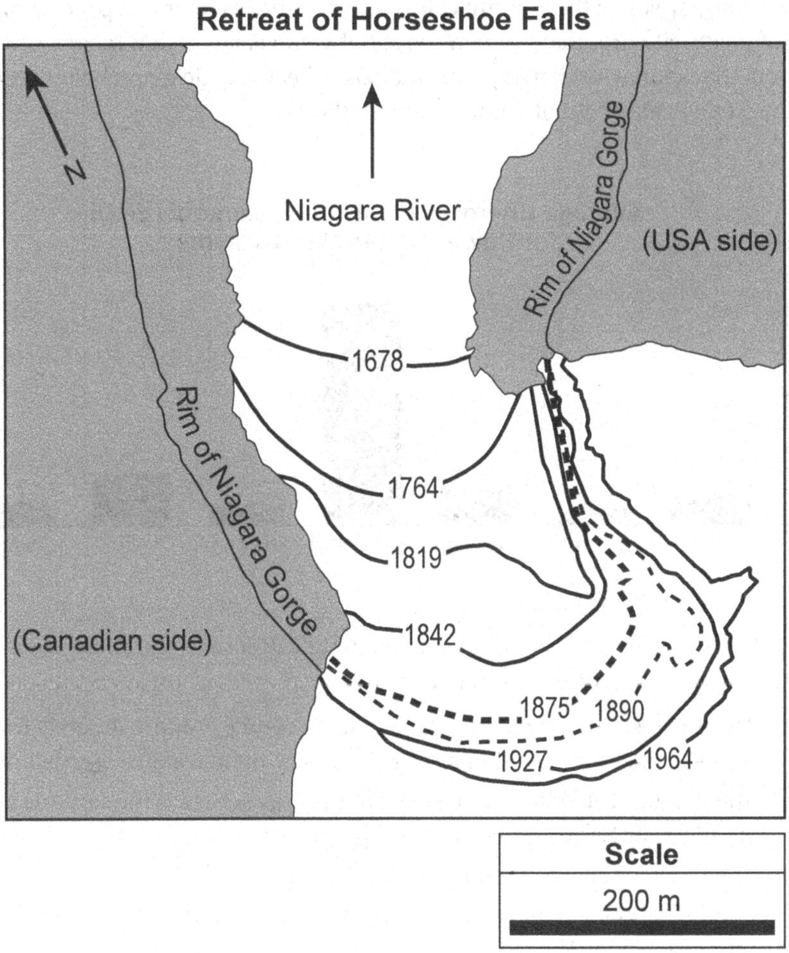

Retreat of Horseshoe Falls

9. Analyze data from the **two** maps to make a claim about how the construction of the hydroelectric power plants and diversion tunnels caused the rate of erosion at Horseshoe Falls to decrease. [1]

____

____

____

No source of electricity is 100% efficient. All sources of electricity require substantial monetary investment to build structures, maintain them, employ workers, and establish a power grid to get electricity to consumers. The energy return on investment calculation (EROI) is a comparison of the amount of money invested in electricity generation to the amount of electricity produced. When the EROI value is large, that means producing energy from that source is relatively easy and cost-effective. The graph below compares EROI values for seven electricity production methods.

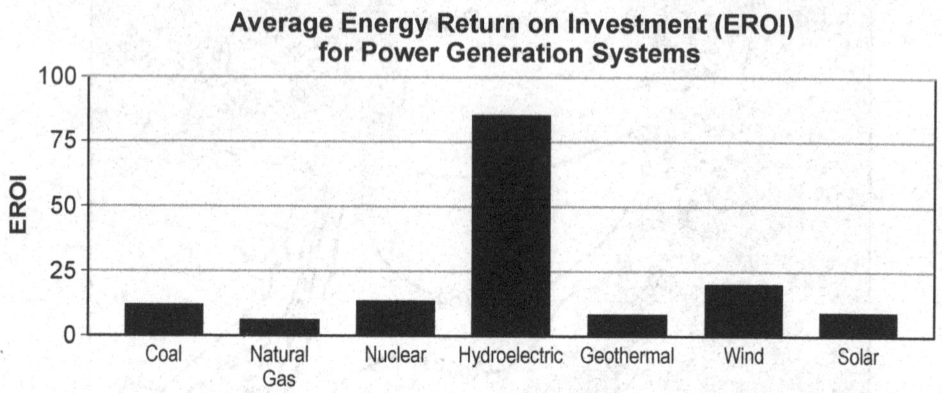

10. Based on the graph, which statement accurately describes the energy return on investment (EROI) for the energy sources?

(1) The greatest EROI for a renewable energy source is approximately five times greater than the greatest EROI for a nonrenewable energy source.

(2) The greatest EROI for a nonrenewable energy source is the same as the smallest EROI for a renewable energy source.

(3) The EROI for coal is the same as the EROI for wind.

(4) The EROI for each renewable energy source is half the EROI for each nonrenewable energy source.

10 ______

Base your answers to questions 11 through 15 on the information below and on your knowledge of Earth and Space Sciences. Some questions may require the use of the ***2024 Edition Reference Tables for Earth and Space Sciences.***

Telescopes and the History of the Universe

Two space telescopes that are providing astronomers with information about the early universe are the Hubble Space Telescope (HST) and the James Webb Space Telescope (JWST).

The HST orbits Earth at a distance of 540 km above the surface and can observe visible, ultraviolet (UV), and infrared (IR) radiation. It contains cameras and spectrographs that break light into the colors of the spectrum for analysis of elements found in stars.

Hubble has determined the age of the universe to a more precise date of 13.8 billion years old and has photographed galaxies in all stages of evolution—even galaxies from the early universe.

A central idea of the Big Bang theory is that the universe expanded rapidly and released energy. Scientists have been gathering evidence for this expansion and energy release since the early 1920s.

The photograph below, from 1964, shows two scientists in New Jersey with a radio telescope aimed towards outer space. This telescope was one of the first to record cosmic microwave background radiation coming from all directions in space.

11. Which explanation about the cosmic microwave background radiation provides evidence for the Big Bang theory?

(1) Since the cosmic microwave background radiation was detected coming from all directions, it must be evidence for multiple explosions occurring at the same time 13.8 billion years ago.

(2) Mathematical calculations showed that if the universe began with a Big Bang, it would have released a huge amount of energy in all directions and would be detected as cosmic microwave background radiation.

(3) Cosmic microwave background radiation is produced by every star and galaxy in the universe, which were created at the same time as the Big Bang.

(4) The presence of cosmic microwave background radiation is evidence for the energy released from dying stars that exploded in a Big Bang and created our present day universe.

11 ______

The only evidence astronomers have for the first stars that formed is the electromagnetic radiation they left behind. Telescopes like Hubble can detect objects billions of light-years away (billions of years back in time).

The graph below shows some information about the first stars that formed after the Big Bang.

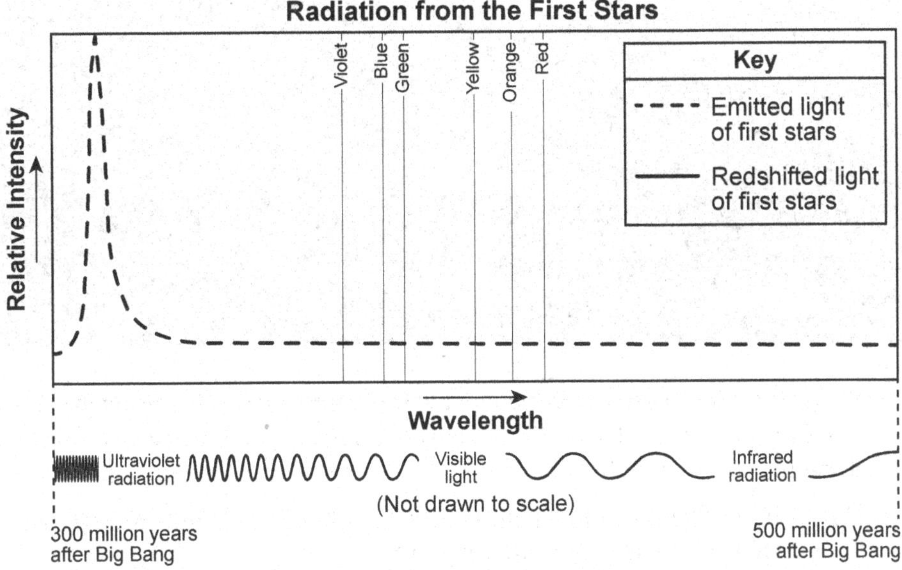

12. Describe how wavelengths of emitted radiation from first stars have changed since the Big Bang and explain how this difference in wavelength is evidence for how the universe changed since the Big Bang. [1]

Difference in wavelength: ____________________

____

How the universe changed: ____________________

____

____

One of the oldest stars ever observed by HST is classified as a giant star called Methuselah in the Milky Way Galaxy. It is believed to be almost as old as the universe itself. Based on Methuselah's composition, scientists do not believe it is one of the first stars.

Digital Image of Methuselah in the Constellation Libra

13. Which statement correctly explains why the composition of Methuselah, a massive star, provides evidence that Methuselah is ***not*** one of the first stars formed during the Big Bang?

(1) Methuselah is composed of 100% helium, but the first stars were composed of 75% hydrogen and 25% iron.

(2) Methuselah is composed of hydrogen, helium, and some trace heavier elements, but the first stars contained a greater percentage of heavier elements.

(3) The first stars were composed of 100% hydrogen, but Methuselah is composed of 50% hydrogen and 50% helium.

(4) The first stars were composed of 75% hydrogen and 25% helium, but Methuselah also contains metals of heavier elements in its core.

13 ______

The model represents the locations of the HST orbit and the Moon's orbit around Earth. The Moon's orbital distance varies from approximately 363,100 km to 405,700 km.

Orbits of HST and the Moon Model

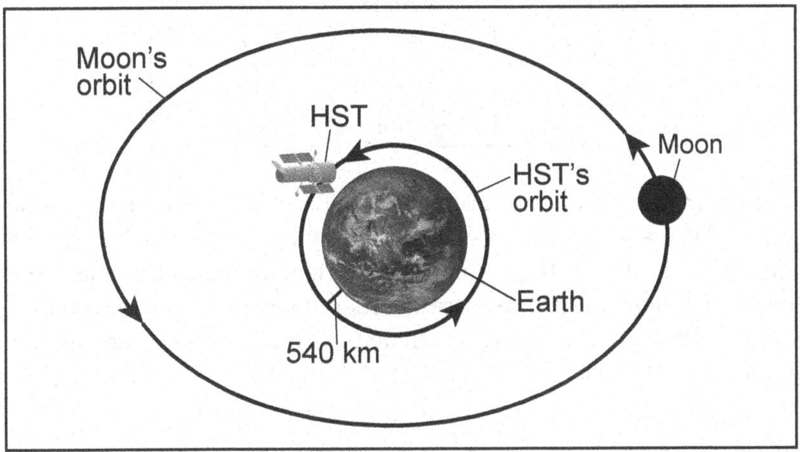

(Not drawn to scale)

The orbital motions of satellites, natural or human-made, are determined by the strength of the gravitational field at the satellite's location. The model below shows the formula used to calculate the strength of Earth's gravitational field for Earth's satellites.

Gravitational Field Model

$$g = \frac{Gm_E}{r^2}$$

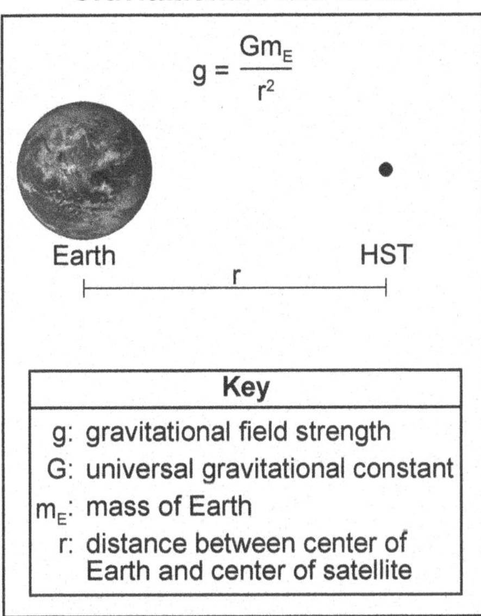

Key
g: gravitational field strength
G: universal gravitational constant
m_E: mass of Earth
r: distance between center of Earth and center of satellite

14. Based on the *Gravitational Field Model,* use the word list to complete the passage by placing the correct terms on the lines below to describe the effects of Earth's gravitational field on the motion of HST and the Moon. [1]

Word List

A	B	C
less	less	less
greater	greater	greater

Compared to the average distance from Earth to the Moon, the distance from Earth to HST is ___A___. As a result, Earth's gravitational field strength at the location of HST is ___B___ than Earth's gravitational field strength at any location of the Moon's orbit. Therefore, in order to remain in a stable orbit, the orbital period of HST must be ___C___ than the orbital period of the Moon.

A: ______________

B: ______________

C: ______________

15. Based on the model and Kepler's Laws, which graph correctly represents the relative orbital speeds of HST and the Moon as the two satellites orbit Earth *twice*?

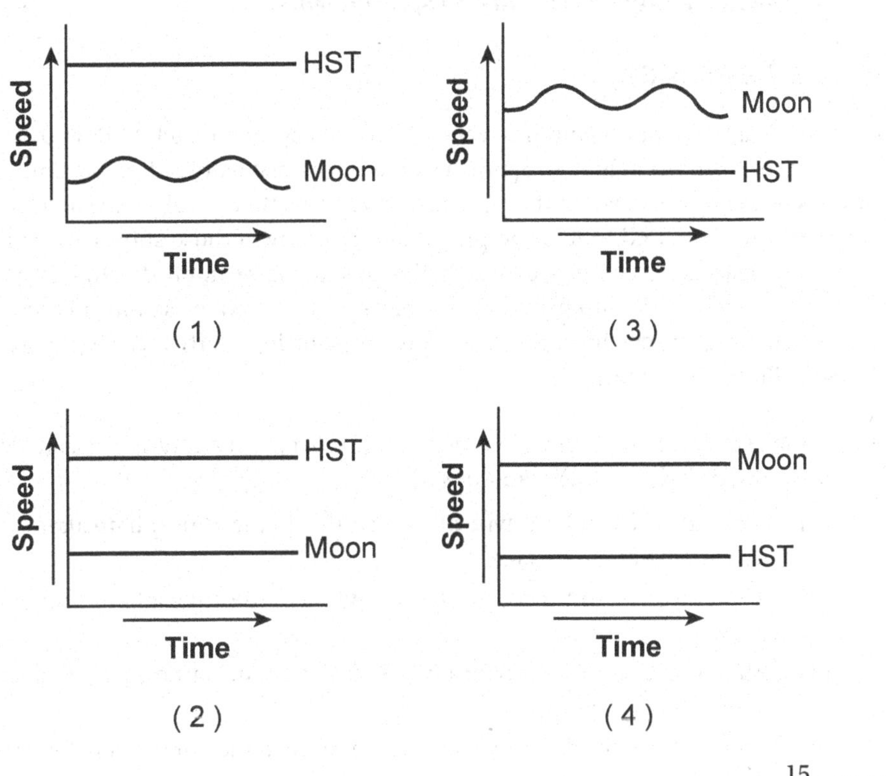

15 ______

Base your answers to questions 16 through 20 on the information below and on your knowledge of Earth and Space Sciences. Some questions may require the use of the ***2024 Edition Reference Tables for Earth and Space Sciences****.*

Horse Latitudes

The historical term "horse latitudes" (areas located at approximately 30°N and 30°S latitude) comes from the legend of Spanish sailing ships that would often become stalled for days or even weeks when they encountered areas that usually experienced calm winds, sunny skies, and little or no precipitation. Many of these ships carried horses to the Americas as part of their cargo. Unable to sail and resupply due to lack of wind, crews often ran low on drinking water. To conserve scarce water, sailors on these ships would sometimes throw the horses they were transporting overboard. Thus, the phrase "horse latitudes" was born.

16. Which statement correctly identifies the air movement patterns associated with the planetary wind "horse latitudes"?

(1) Air at these latitudes is typically ascending in the atmosphere after converging at the surface.

(2) Air at these latitudes is typically ascending in the atmosphere after diverging at the surface.

(3) Air at these latitudes is typically descending in the atmosphere before converging at the surface.

(4) Air at these latitudes is typically descending in the atmosphere before diverging at the surface.

16 ______

Map of Air Mass Source Regions

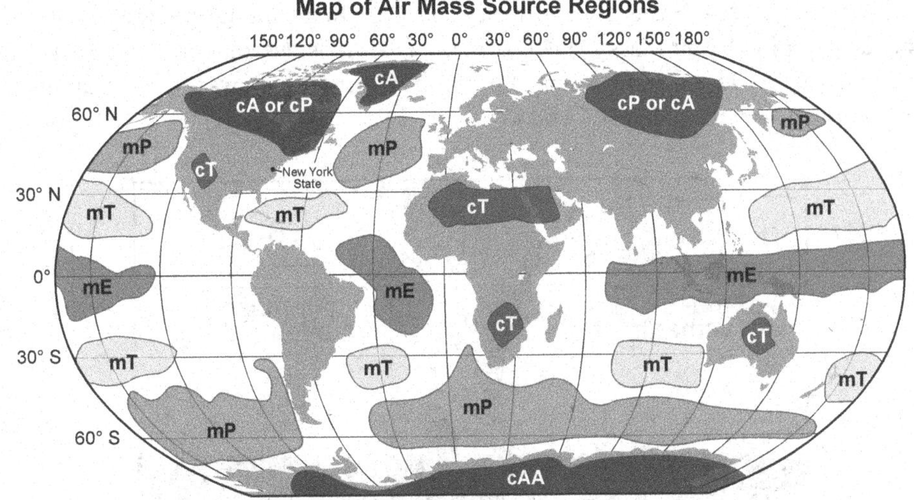

Data Table: Air Masses

Air Mass Name	Air Mass Code
Continental Polar	cP
Continental Tropical	cT
Maritime Polar	mP
Maritime Tropical	mT
Maritime Equatorial	mE
Continental Arctic	cA
Continental Antarctic	cAA

17. Which winds drive the movement of a continental tropical air mass across Earth's surface away from the northern "horse latitudes" toward 60°N?

(1) Northeast trade winds move cT air masses from northeast to southwest.

(2) Prevailing westerly winds move cT air masses from northeast to southwest.

(3) Northeast trade winds move cT air masses from southwest to northeast.

(4) Prevailing westerly winds move cT air masses from southwest to northeast.

17 ______

Utica, New York, is located at approximately 43°N. In the model below, Utica is influenced by a cold, dry air mass (*A*). A warm, moist air mass (*B*) that originated along the east coast of the United States enters the region. The model shows these two air masses interacting near a residence in Utica.

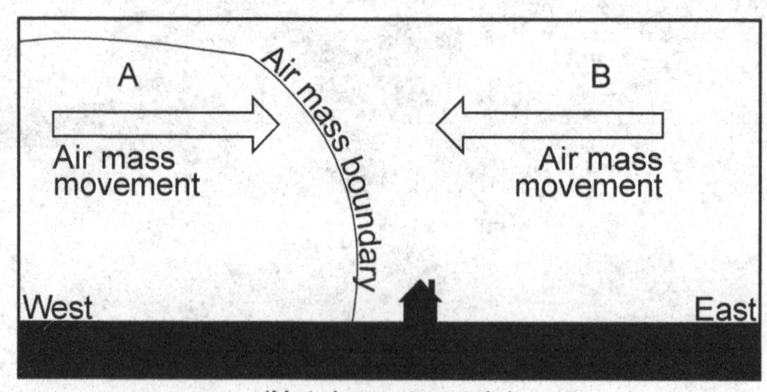

(Not drawn to scale)

18. Identify each air mass using its two-letter code. Also, place a checkmark (✓) in the box that indicates the resulting change in weather conditions in the next few hours that would most likely be experienced by an observer who resides in the house shown in the model. [1]

Code for Air Mass *A*: ________

Code for Air Mass *B*: ________

Weather Conditions	Increases	Decreases
cloud cover		
chance of precipitation		
air temperature		

The Bermuda high pressure system forms in the "horse latitude" region of the North Atlantic Ocean. Air circulation around the margins of the Bermuda high is known as "steering flow" and can influence a hurricane's storm track. Meteorologists analyze the position and strength of Bermuda high pressure systems to help make predictions about hurricane storm tracks and determinations about where hurricanes will make landfall. The maps below show some information about the effects of Bermuda highs.

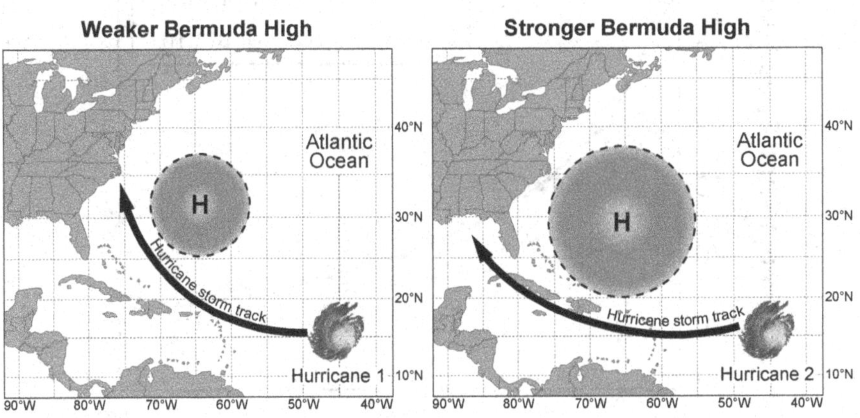

19. Which table correctly compares the characteristics of Hurricane 2 to Hurricane 1 and its associated Bermuda high?

(1)

Steering Flow	Storm Track	Bermuda High
farther south, then west	west, then north	clockwise and strong

(2)

Steering Flow	Storm Track	Bermuda High
farther south, then west	east, then north	clockwise and weak

(3)

Steering Flow	Storm Track	Bermuda High
farther north, then west	west, then north	counterclockwise and strong

(4)

Steering Flow	Storm Track	Bermuda High
farther north, then west	east, then north	counterclockwise and weak

19

The graph below shows some information about hurricanes and tropical storms.

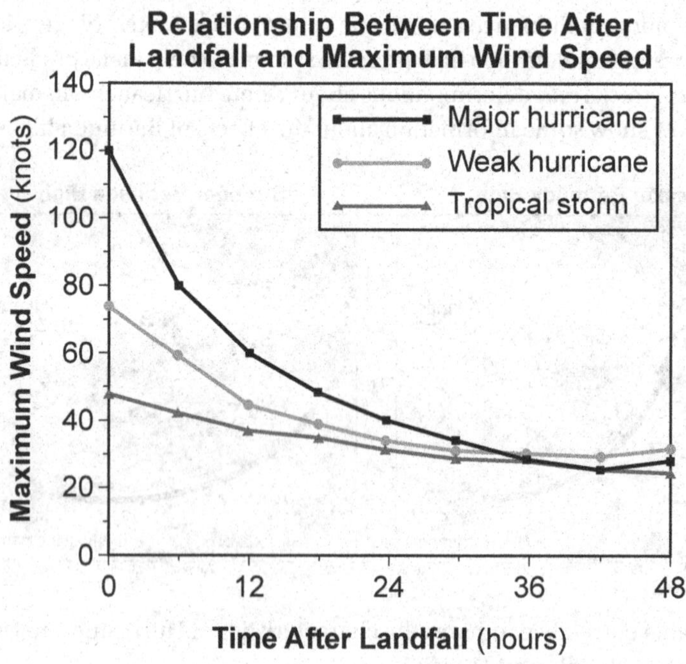

20. Compare the atmospheric pressure at 12 hours after landfall to the atmospheric pressure at 24 hours after landfall using the wind speed and landfall data shown in the graph to justify your response. [1]

____

____

____

Base your answers to questions 21 through 26 on the information below and on your knowledge of Earth and Space Sciences. Some questions may require the use of the ***2024 Edition Reference Tables for Earth and Space Sciences****.*

How Did Earth Get Its Water?

Scientists are unsure about where and how Earth got its water. Early Earth lacked an atmosphere and was very hot, so water would not have formed in such an environment. One theory is that water was brought to Earth when comets collided with Earth, since comets are made of frozen water, gases, and dust.

Another theory centers around asteroids. Earth is believed to have formed from the clumping together of space debris. These early solar system materials contained the elements hydrogen and oxygen. The chemical proportions of the water found on asteroids are similar, though different, from Earth's water.

The most recent discoveries have come from studies of dust particles brought back to Earth from the Hokawa asteroid. The particles were found to have water and an oxygen-hydrogen ion. The source of the hydrogen is believed to be from solar wind—hydrogen ions ejected from the Sun streaming through space and lodging onto the surface of dust particles. These particles have been falling to Earth since the solar system formed. The space dust is believed to have provided about 50% of Earth's water.

The graph below shows the relative deuterium-to-hydrogen ratio (D/H) in water found in solar system objects. Diamond symbols are measurements obtained by satellites. Dots are estimated values. Deuterium is a "heavier" form of hydrogen. This ratio is an important factor in determining where in the solar system an object formed and how asteroids, comets, and the solar wind contributed to the water in Earth's oceans.

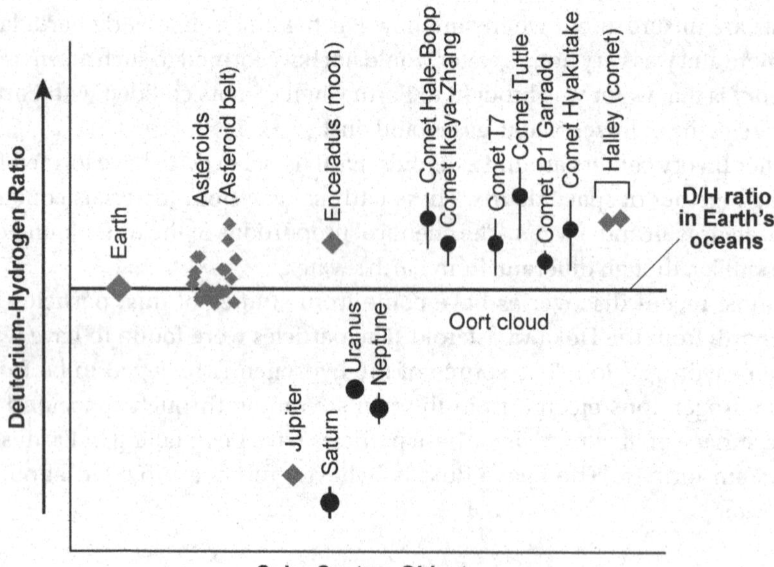

21. A student makes a claim that the water on our planet is most similar to the water found on comets based on its D/H ratio, so comets must be a major source of water on early Earth. Which statement, based on evidence from the graph, would *refute* this claim?

(1) The source of Earth's water is the four largest planets in our solar system because these planets have water with higher D/H ratios than Earth's oceans.

(2) The water on all the comets has the same D/H ratios except Halley, so Halley is not a source of water for Earth's early oceans.

(3) Many asteroids have water that is very close in D/H ratios to Earth's oceans, so asteroids could be a source of water on early Earth.

(4) Water on Enceladus has a similar D/H ratio to comets, so Enceladus could also be a source of water for Earth.

21 ______

Earth's first significant atmosphere was formed from the release of gases from volcanic eruptions, when the mantle is believed to have been molten. Gases from the interior were released and accumulated into a gaseous surface layer trapped by gravitational forces. Solar wind from the Sun swept the light gases (hydrogen and helium) away, leaving behind heavier gases in the early atmosphere.

The models below show gases entering Earth's atmosphere at two different times.

Early Earth Volcanic Outgassing Model

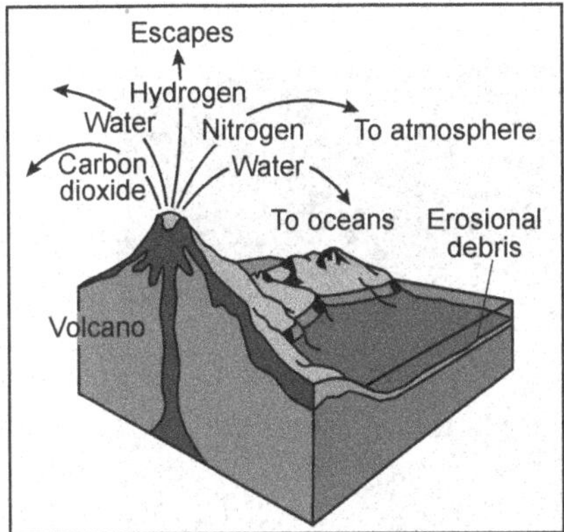

Present Day Volcanic Outgassing Model

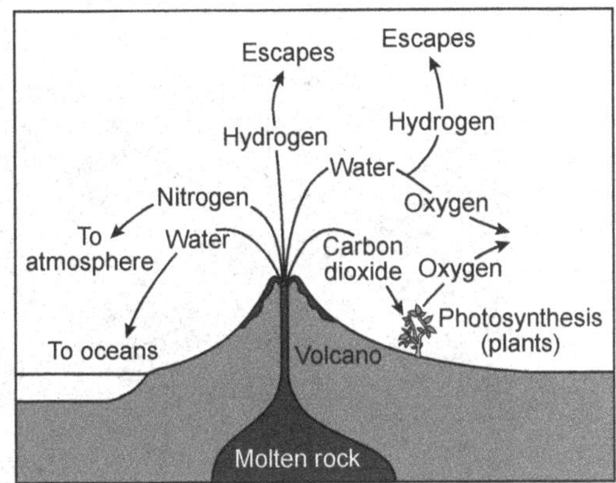

22. Which statement is correctly supported by evidence from the two models showing volcanic outgassing?

(1) All gases emitted from early Earth volcanoes escaped to outer space.

(2) The outgassing of water from Earth's interior contributed to the development of Earth's early oceans.

(3) Earth's early atmosphere contained significant amounts of hydrogen and helium as a result of volcanic outgassing.

(4) Present-day volcanoes directly release oxygen that is used by marine and land plants, which in turn release carbon dioxide.

22 ______

The water in Earth's early oceans provided the necessary environment for life about 3 to 3.5 billion years ago. Organisms such as stromatolites were the earliest example of life on Earth and were capable of photosynthesis.

From about two billion years ago to 500 million years ago, Earth's atmosphere changed significantly, driven by an increase of many different life forms. Living and fossilized stromatolites can still be found today in Australia.

2.7-Billion-Year-Old Fossilized Stromatolite – Australia

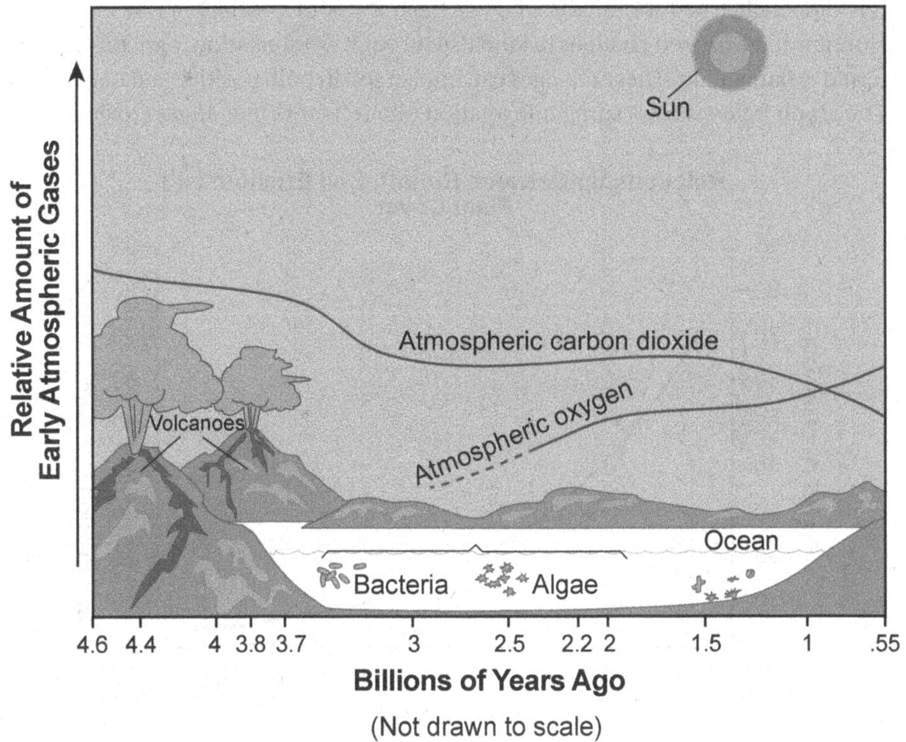

23. Construct an argument, based on evidence in the model, about how stromatolites in Earth's oceans caused Earth's early atmosphere to evolve. [1]

____

____

____

____

As life on Earth evolved and expanded onto land, fresh water became a vital resource. This is especially true for humans who use fresh water in a variety of ways.

Humans have caused changes in land use through deforestation, agricultural expansion, and urbanization. These changes can impact soil in both positive and negative ways.

The graph below shows some information about factors that affect erosion.

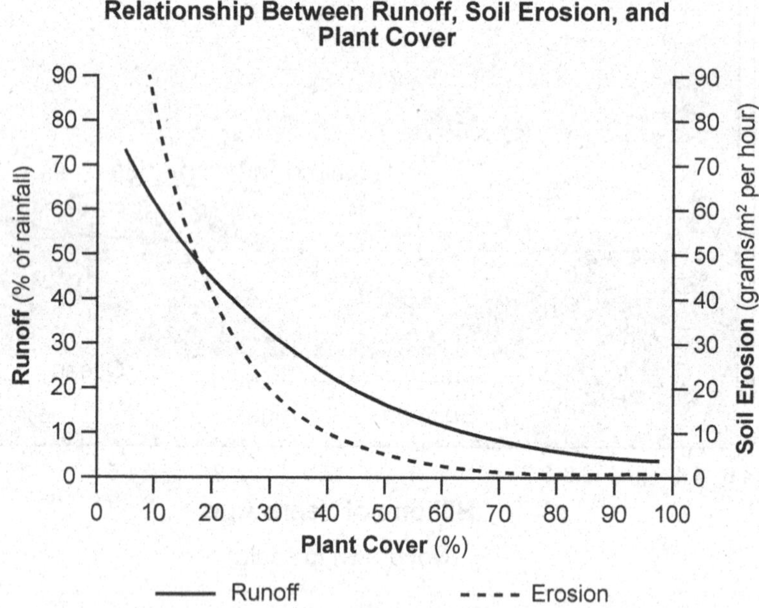

24. Use the graph to complete the claims that describe how changes in the percent of plant cover on a land surface cause changes to runoff and the amount of soil eroded during a rain event. [1]

Claim #1: As the percent of plant cover (increases *or* decreases) runoff

circle one

will ________________________________.

Claim #2: As the percent of plant cover (increases *or* decreases) the amount of

circle one

soil eroded during a rain event will ____________________.

In areas where human population density is high, managing fresh water availability is essential.

A large, underground store of water is called an aquifer. One of the most important aquifers in New York State is beneath Long Island and supplies 400 million gallons of water per day to over 2.8 million people.

The cross section below represents water cycle processes for the Long Island area. The three aquifers (Upper Glacial, Magothy, and Lloyd) and sediment or rock layers that together make up the larger Long Island aquifer region are shown.

Water Cycle and the Long Island Aquifer Model

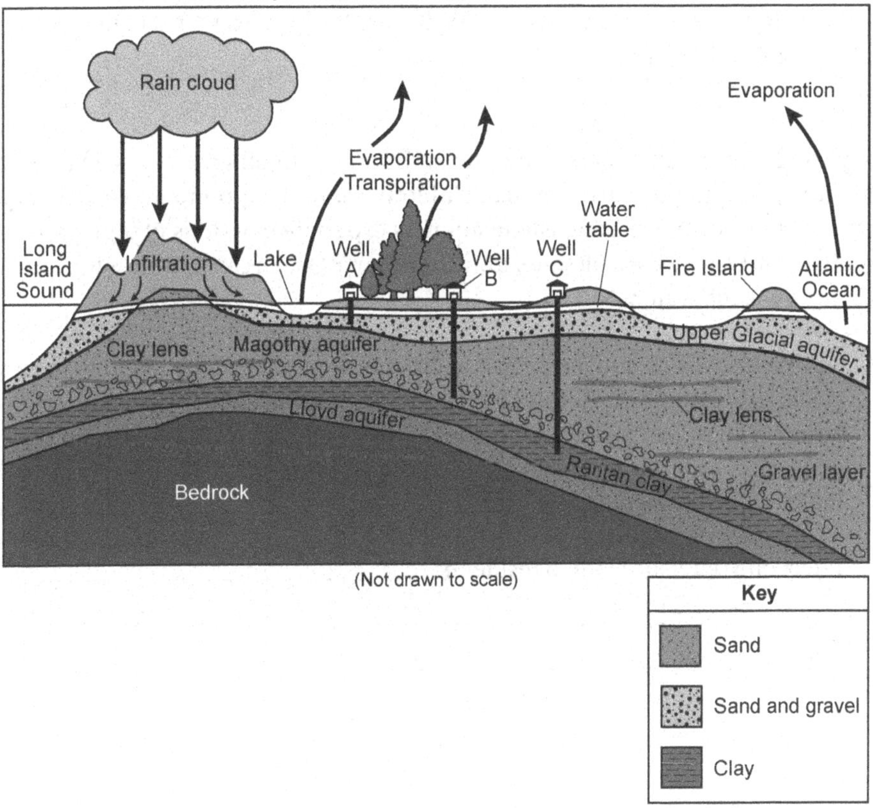

(Not drawn to scale)

Key

Sand

Sand and gravel

Clay

25. A groundwater investigation is planned for water Well *B* and Well *C* from the cross section. A scientist believes that Well *B* will supply water at a faster rate than Well *C*. Which statement most likely explains why the scientist is correct?

(1) The sediments at the base of Well *B* are less porous with a higher permeability than Well *C*.

(2) The sediments at the base of Well *B* are more porous with a higher permeability than Well *C*.

(3) The sediments at the base of Well *C* are more porous with a lower permeability than Well *B*.

(4) The sediments at the base of Well *C* are less porous with a higher permeability than Well *B*.

25 _____

A period of drought occurs when the amount of rainfall an area receives and the amount of precipitation that eventually infiltrates into the ground to reach an aquifer is greatly reduced. When Long Island aquifers experience periods of drought or over-pumping from water wells, salt water from the Atlantic Ocean can intrude into the aquifer, polluting the fresh water.

26. Describe how a prolonged period of drought would affect the depth of the water table. Also, describe how this change would affect residents supplied by water from Well *A* more than residents supplied by water from Well *B*. [1]

Water table depth: ____

Effect on Well *A* and Well *B* residents: ______________________________

____

____

Base your answers to questions 27 through 30 on the information below and on your knowledge of Earth and Space Sciences. Some questions may require the use of the ***2024 Edition Reference Tables for Earth and Space Sciences.***

Northern New York's Pyrites Complex

The Pyrites Complex, located in New York State's Northwest Lowlands, represents a 1.35- to 1.21-billion-year-old fragment of ocean crust and upper mantle rocks. Most of the rocks in the Northwest Lowlands are metamorphosed sedimentary and igneous rocks. The map below shows the location of the Pyrites Complex within the Northwest Lowlands. Bodies of water are shown in black.

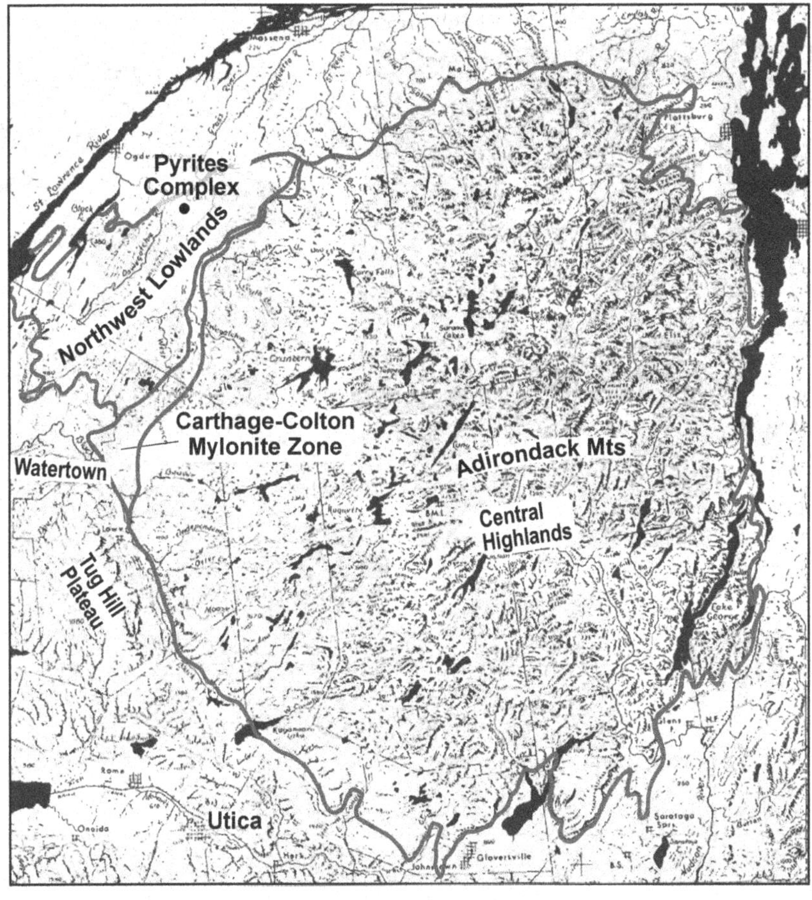

Models *A*, *B*, *C*, and *D* represent the formation of the Pyrites Complex. Arrows indicate the direction of plate motion.

Formation of the Pyrites Complex

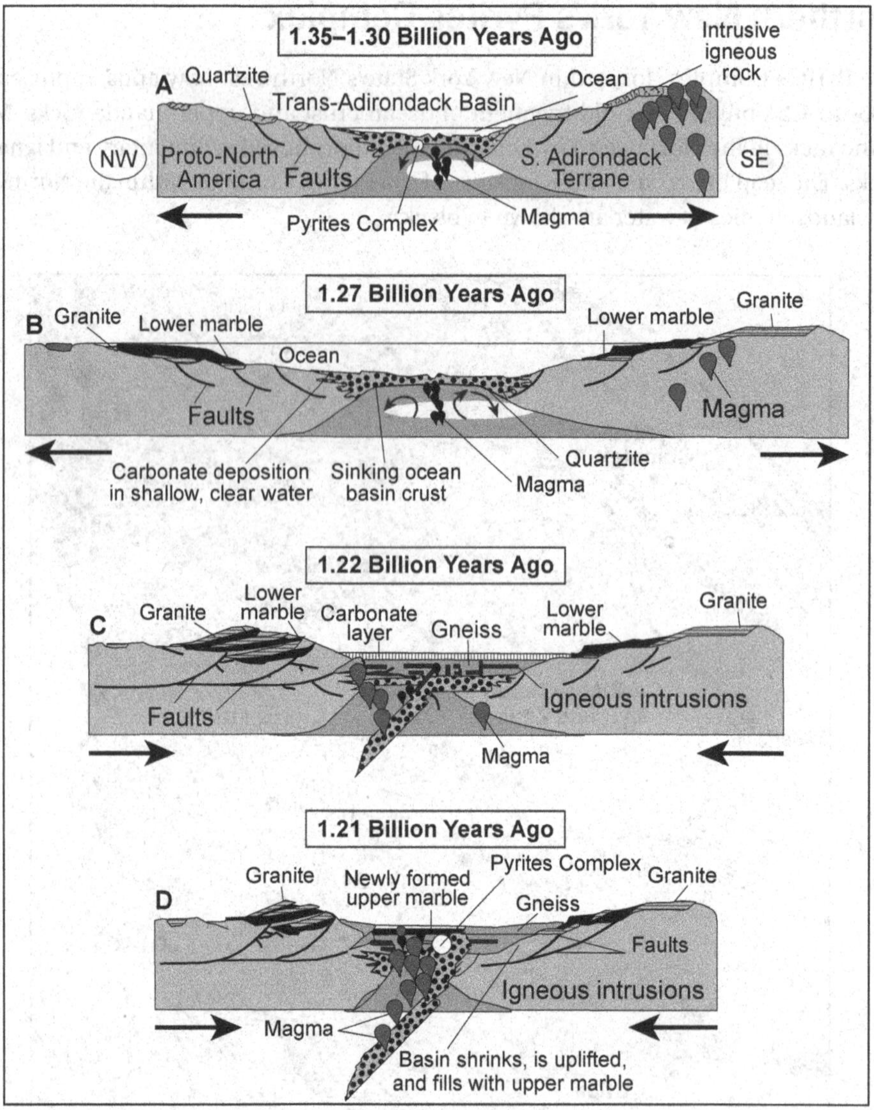

(Not drawn to scale)

27. Use **one** of the *Formation of the Pyrites Complex* models to identify **one** geologic process responsible for the cycling of matter that contributed to the formation of the Pyrites Complex. Include the letter of the model in your answer. [1]

Letter of model: ___________

Geologic process: ____

____

28. A student made a chart of constructive and destructive geologic processes from information shown in the *Formation of the Pyrites Complex* models.

Constructive	Destructive
Model B: new ocean crust formed	**Model C:** subduction of oceanic crust
Model A: intrusive igneous rock rises to the surface	**Model D:** ocean basin shrinks
Model B: deposition of sediment along ocean basin	**Model D:** marble is formed from metamorphism of limestone

Which table below shows an additional process that is classified correctly and could be added to the table above?

(1)

Constructive
Model A: Trans-Adirondack Basin shrinks

(2)

Destructive
Model B: faults first appear along either side of the ocean basin

(3)

Constructive
Model C: granite and lower marble are uplifted

(4)

Destructive
Model D: basin fills with upper marble

28 ______

The flow chart below shows four geologic stages that led to the formation of the Pyrites Complex. The second geologic stage was intentionally left blank.

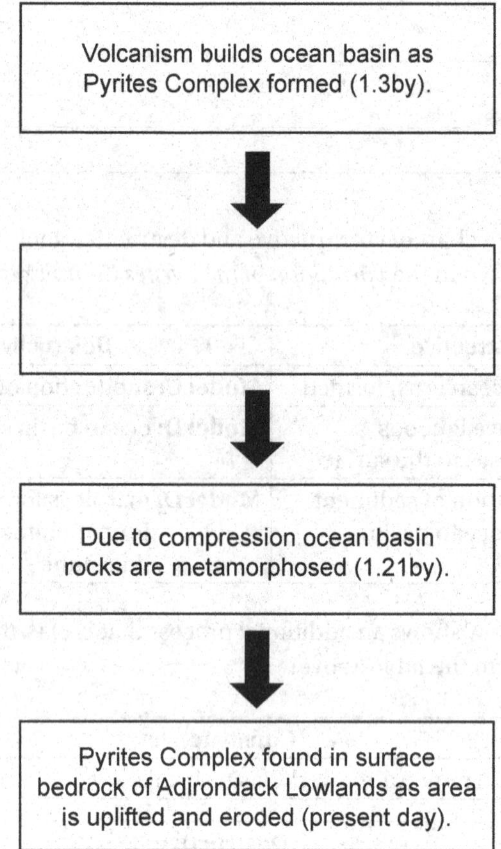

29. Based on evidence from the *Formation of the Pyrites Complex* models, which geologic process correctly identifies the missing stage in the flow chart above?

(1) Quartzite is formed from lower marble.

(2) The ocean floor continues to form and widens.

(3) Faults move granite beneath the ocean basin.

(4) Divergence causes the ocean to become narrower and shallower.

29 ______

The model below pairs tectonic locations with metal deposits that form within them.

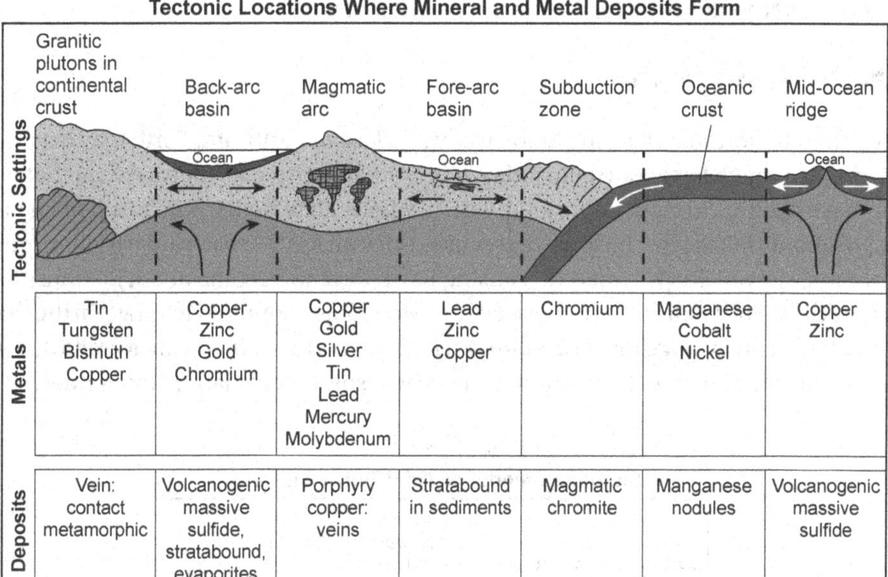

30. Based on the model above, identify the two tectonic settings in which pyrite, a sulfide-based deposit, forms in association with zinc and copper.

(1) Back-arc basin and mid-ocean ridge

(2) Fore-arc basin and magmatic arc

(3) Mid-ocean ridge and granitic plutons in continental crust

(4) Subduction zone and fore-arc basin

30 ______

Base your answers to questions 31 through 35 on the information below and on your knowledge of Earth and Space Sciences. Some questions may require the use of the ***2024 Edition Reference Tables for Earth and Space Sciences.***

Earth-Moon History

About four billion years ago, the Moon was very close to Earth, and Earth may have been rotating at a much faster rate than today. It is estimated that each Earth day, at that time, may have been six to eight hours long, which, when calculated, would result in an Earth year of about 1400 days. The force that causes tides on Earth creates tidal friction which, over time, slowed Earth's rotation, causing Earth days to become about 24 hours long. As Earth's rate of rotation has decreased, the Moon has continued to move farther away from Earth at an average rate of 3.8 cm/yr. The Moon orbits Earth at an average distance of 384,400 km. The model and graph below show some information about tides.

Simplified Model of Tidal Bulge on Earth

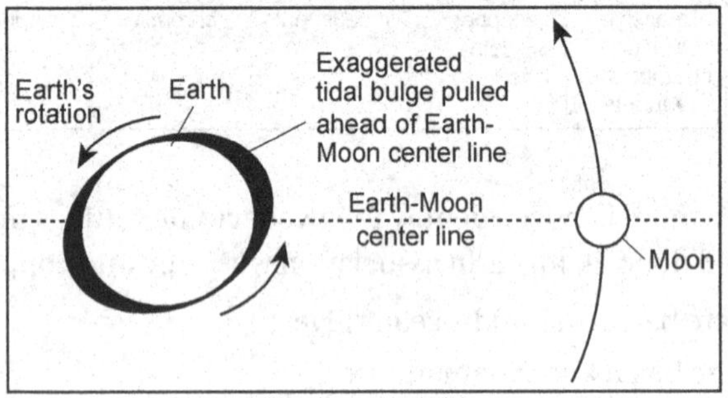

(Not drawn to scale)

Observed Tides at Montauk Harbor, NY

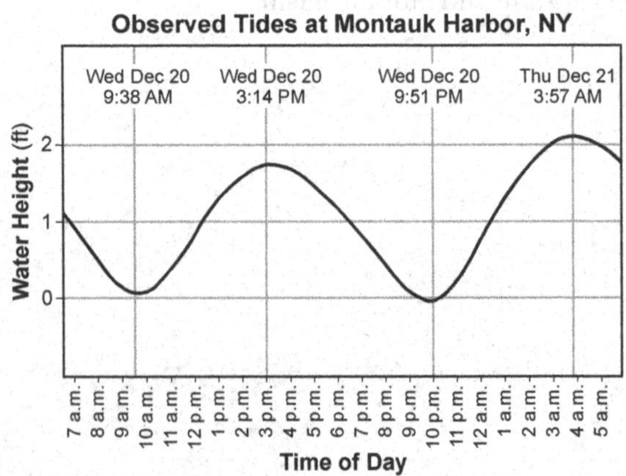

31. Which row in the table below correctly identifies the evidence and explanation for the observed cyclic change in tides at Montauk Harbor?

Row	Evidence	Explanation
(1)	change of about two feet in height of water	The Moon's rotation as it orbits Earth causes water level changes.
(2)	change of about two feet in height of water	Earth's revolution around the Sun causes water level changes.
(3)	high tides approximately 12 hours apart	Earth's rotation results in a tidal bulge at a location approximately twice per day.
(4)	high tides approximately 12 hours apart	The Moon's revolution around Earth results in a tidal bulge at a location approximately twice per day.

31 ______

32. Determine, in centimeters, how much farther away from Earth the Moon's average orbit will be 1000 years from the present day, and explain why this change in distance will *not* significantly affect the Moon's period of revolution around Earth. [1]

Distance: ______________________ **cm**

Explanation: ____

____

The table below shows some information about the Sun and the Moon.

Sun and Moon Data

	Sun at Perihelion	Sun at Aphelion	Moon at Perigee	Moon at Apogee
Image of Apparent Size as Viewed from Earth				
Approximate Distance from Earth (km)	147,100,000	152,100,000	363,300	405,500
Approximate Measured Angular Diameter (°)	0.54	0.52	0.56	0.49

33. Construct an explanation to support the claim that in about 700 million years, a total solar eclipse on Earth's surface will be impossible. [1]

____

____

____

The measured angular diameter of a celestial body is the angle that the diameter of the body makes as seen from Earth.

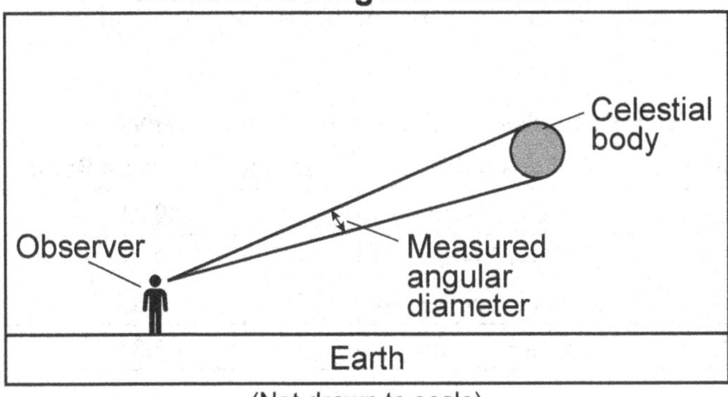

(Not drawn to scale)

34. Based on the information in the table, which lettered pair correctly completes the passage below?

The change in the angular diameter of the Moon during the Moon's orbit is ___A___ than the change in the angular diameter of the Sun during Earth's orbit because the Moon's orbit is inferred to be ___B___ elliptical than Earth's orbit.

(1)	(2)	(3)	(4)
A: smaller	A: greater	A: smaller	A: greater
B: more	B: more	B: less	B: less

34 _____

Astronomers theorize that the Moon formed as a result of a collision between Earth and another planet. The debris created during this collision circulated around Earth and formed the Moon. Evidence for the age of the Moon is found in zircon, a mineral found in lunar rocks. This mineral contains uranium-238.

The table below shows some information about four rock samples taken from the lunar surface.

Lunar Samples and Percent of Uranium Remaining

Sample	Location	Percent of Uranium Remaining
A	Highland Crust	49.31
B	Lunar Mare	79.25
C	Oceanus Procellarum	73.33
D	Imbrium Basin	61.83

35. Using the table, identify the letter of the sample that provides evidence to support the claim that the Moon is approximately 4.53 billion years old. Explain how this evidence supports the claim. [1]

Sample: ____________________

Explanation: ____

____

____

Base your answers to questions 36 through 40 on the information below and on your knowledge of Earth and Space Sciences. Some questions may require the use of the ***2024 Edition Reference Tables for Earth and Space Sciences.***

Dams and Flood Mitigation

Dams and other structures are often put in place in river systems to help reduce life and property losses to residents living near rivers due to flooding. Dams are also built to generate electricity.

The graph below shows river flow rates on the Kootenai River in Montana in cubic meters per second (m^3/s).

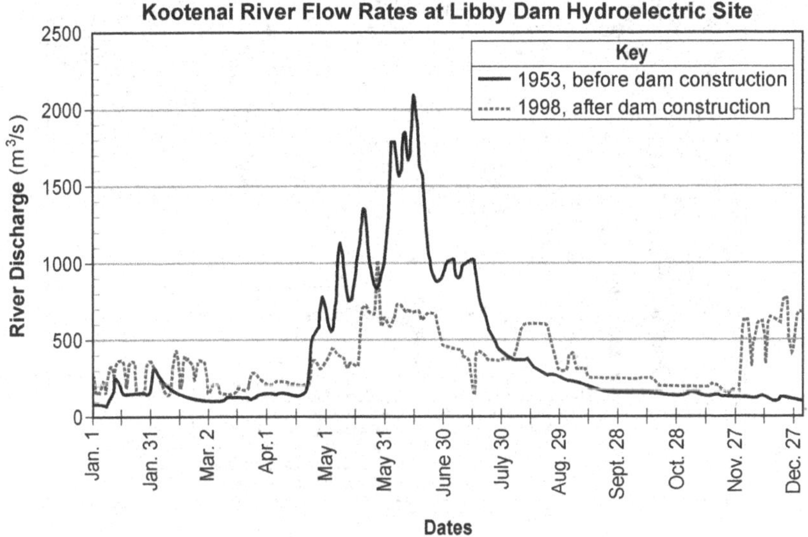

36. Which claim identifies a benefit of developing and managing the Libby Dam Hydroelectric Project after its construction?

(1) The river flow rates have increased for the entire year, increasing the amount of electricity generated.

(2) The river flow rates have decreased during the winter months, increasing the amount of electricity generated.

(3) After dam construction, the river flow rates changed less throughout the year and allowed for more control of electric generation.

(4) After dam construction, the river flow rates changed more throughout the year and allowed for more control of electric generation.

36 ______

The model below shows some information about a river system.

River Ecosystem with a Dam

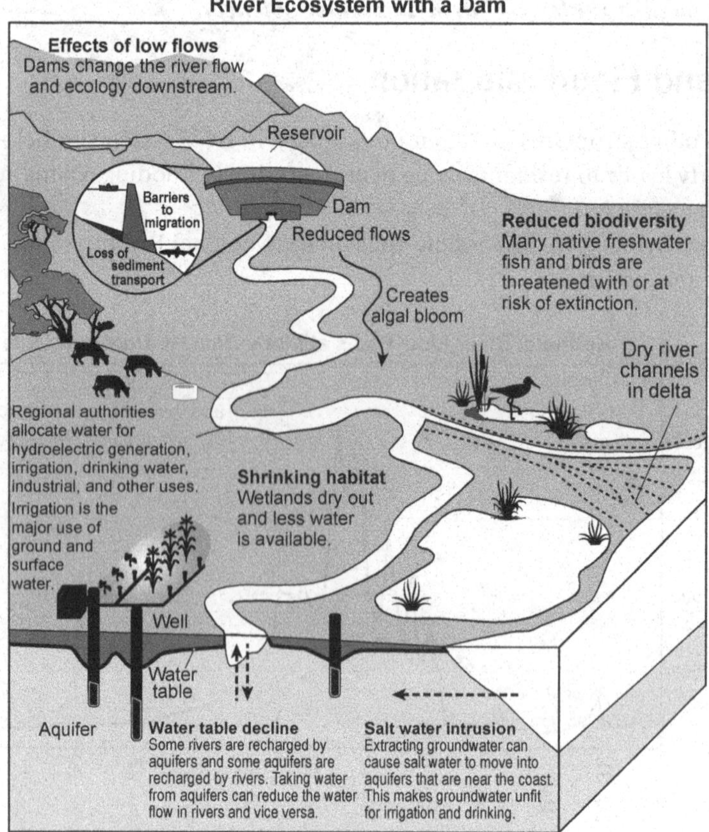

(Not drawn to scale)

37. Which statement correctly identifies a relationship between the hydrosphere and the biosphere in a river system with *reduced* water flow below a dam?

(1) Lower water levels below the dam will lower the water table and will increase biodiversity of native fish and birds.

(2) The water table will rise and provide a benefit to agricultural areas, industrial areas, and wetlands.

(3) Reduced oxygen levels in the stream will create algal blooms, which improve habitat for fish.

(4) A decrease in water flow promotes salt water intrusion, which is harmful to certain plants and animals.

37 ______

38. Use the model to construct a claim about the amount of coastal sediment deposited by a river system before and after a dam is constructed. [1]

____

____

____

The diagrams below show a river system before and after being dammed to create a reservoir. Arrows indicate direction of flow of the river from upstream to downstream.

Free-Flowing River

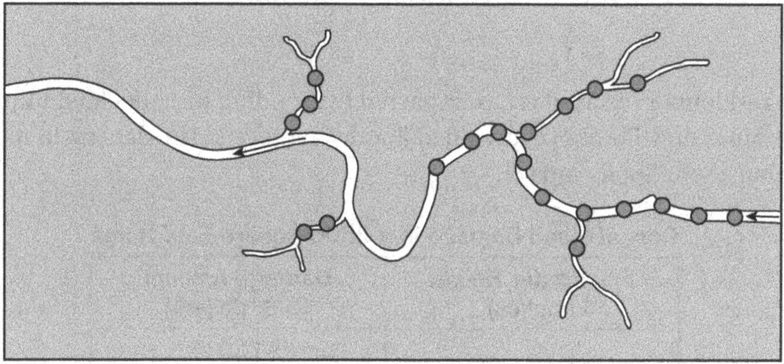

River with Dam and Reservoir

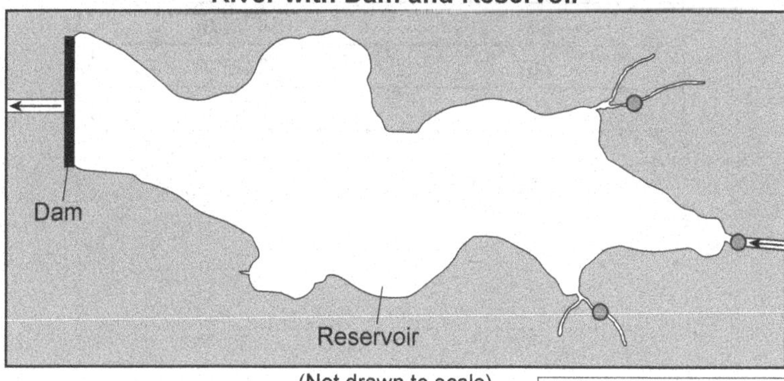

(Not drawn to scale)

Key
● Fish spawning and rearing habitat

39. Based on the *River Ecosystem with a Dam* model and the diagrams, which evidence-based statement correctly identifies how the availability of fish-spawning habitats has influenced commercial fishing above the dam?

(1) Dams increase water levels upstream, decreasing fish spawning and decreasing commercial fishing opportunities.

(2) Dams increase dissolved oxygen downstream, decreasing water quality and decreasing commercial fishing opportunities.

(3) Dams decrease water levels upstream, increasing fish spawning and decreasing commercial fishing opportunities.

(4) Dams decrease dissolved oxygen downstream, increasing water quality and increasing commercial fishing opportunities.

39 ______

Some residents in coastal areas impacted by flooding have decided to place their existing homes on stilts above identified floodwater levels. The data table shows some information about flood damage.

Cost of Flood Damage in a 2500-Square-Foot Home

Floodwater Height (inches)	Damage Amount (U.S. dollars)
6"	$52,037
12"	$72,162
24"	$87,326
36"	$94,538
48"	$103,355

Home Being Raised Onto Stilts

40. Use the *Cost of Flood Damage in a 2500-Square-Foot Home* table to calculate how much a homeowner could potentially save after spending $56,250 to raise their 2500-square-foot home to avoid being impacted by a 24-inch flood. [1]

Amount saved: $ ____________

Base your answers to questions 41 through 45 on the information below and on your knowledge of Earth and Space Sciences. Some questions may require the use of the ***2024 Edition Reference Tables for Earth and Space Sciences.***

New York State Fossils of Note

The New York State fossil shown below is a Silurian-age eurypterid called *Eurypterus remipes.*

This fossil was originally believed to be a catfish before being correctly described as an arthropod—an invertebrate with a segmented exoskeleton and hinged appendages. Some eurypterid appendages were used for walking, while other appendages were used for swimming.

New York State Fossil
Eurypterus remipes

The model below is a family-level evolutionary tree (cladogram) for various eurypterids. The bars represent known temporal ranges for each family. Some eurypterid silhouettes are shown to scale with lines indicating their temporal placement in each family.

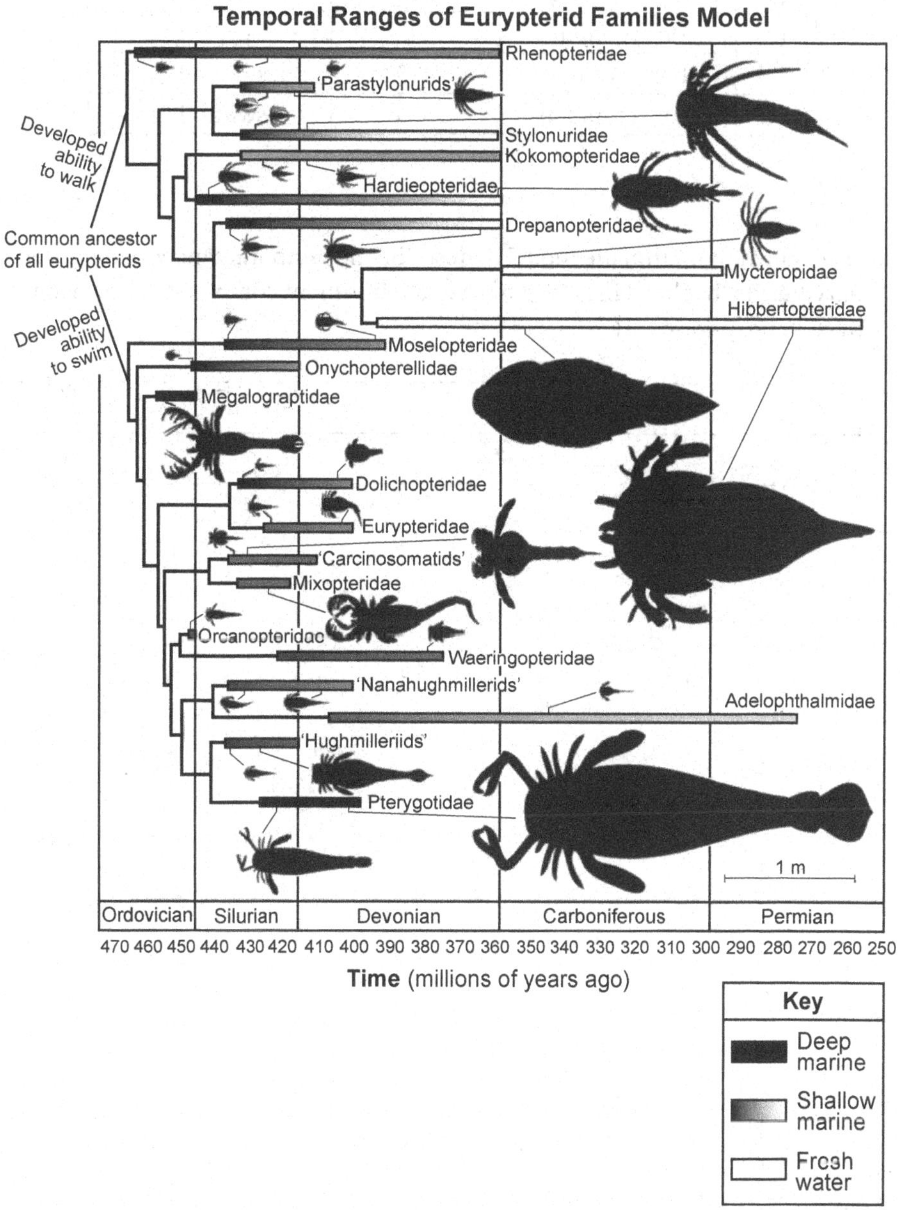

Temporal Ranges of Eurypterid Families Model

Time (millions of years ago)

Key
Deep marine
Shallow marine
Fresh water

41. Which row in the table correctly identifies the family of *Eurypterus remipes*, the age of the fossil, and the type of movement its appendages were used for?

Row	Family	Age (mya)	Type of Movement
(1)	Rhenopteridae	350	walking
(2)	Pterygotidae	330	walking
(3)	Eurypteridae	420	swimming
(4)	Mycteropidae	355	swimming

41 ______

42. Use information from the model to describe the temporal ranges of fossilized eurypterids that lived in a deep marine environment compared to those that lived in fresh water. [1]

____

____

____

43. Based on evidence in the model, which table correctly completes the passage below?

Evidence of the mass extinction of __A__ eurypterid families can be inferred from the model. This extinction event occurred at the end of the __B__ period. After this event, the majority of remaining eurypterid families lived in a __C__ environment.

(1)

A	walking
B	Silurian
C	fresh water

(2)

A	walking
B	Devonian
C	fresh water

(3)

A	swimming
B	Devonian
C	deep marine

(4)

A	swimming
B	Silurian
C	deep marine

43 ______

One of the world's oldest fossil forests was found in the 1850s in Gilboa, New York. Preserved plant remains were unearthed at the Riverside Quarry dated to about 390 million years ago in the Devonian Period. This was a time when land plants were evolving into forest ecosystems.

Four hundred and eighty-six objects (root mounds, horizontal stems, and plant fragments) were identified in the rock, including two types of trees. *Eospermatopteris* trees were found and are related to ferns. Aneurophytalean progymnosperms called *Tetraxylopteris* also were found. *Tetraxylopteris* grew underground but sprouted above ground. These same fossil trees have also been found in Venezuela and Morocco.

The map below shows the location of landmasses when the first ancient forests existed during the Devonian Period.

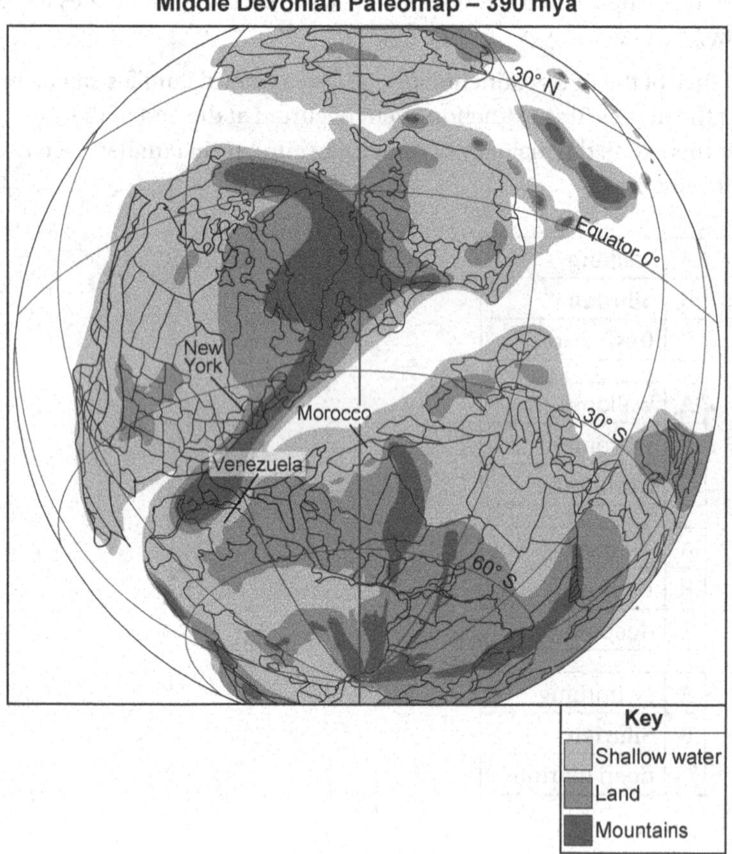

44. Which evidence from the paleomap could be used to support an explanation for the presence of similar ancient plant and tree fossils found in Gilboa, NY; Morocco; and Venezuela?

(1) These three landmasses were located next to each other during the Devonian Period.

(2) These three landmasses were located in a similar climate zone that allowed ancient forests to survive during the Devonian Period.

(3) All of New York, Morocco, and Venezuela were northern hemisphere landmasses located above water during the Devonian Period.

(4) New York, Morocco, and Venezuela had tropical ecosystems that allowed for ancient forests to develop in mountainous areas.

44 ______

The image below shows the fossils found in sandstones and mudstones near Gilboa, New York, and an artist's reconstruction of this type of tree typically found in the Devonian forest.

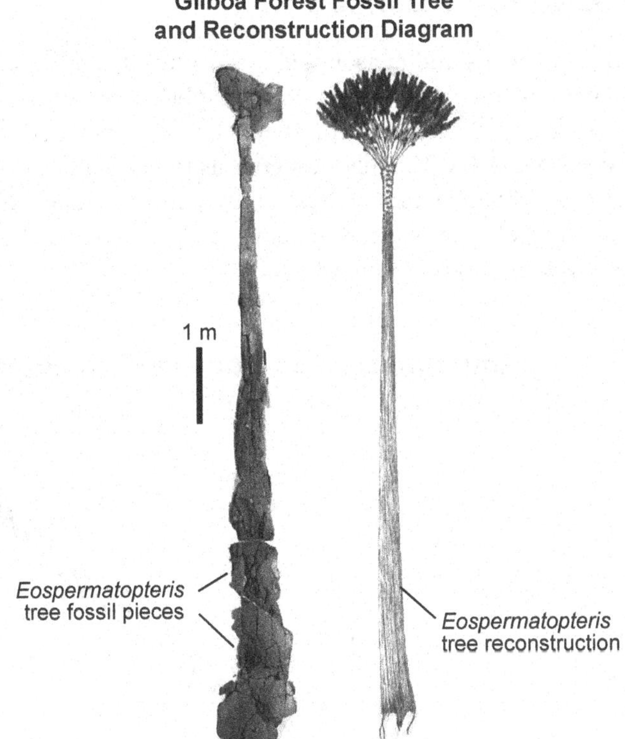

45. Based on the diagram, which row in the table below provides the correct evidence that accurately describes the height (in meters) and environment of formation for the *Eospermatopteris* tree fossil?

Row	Height of Tree (meters)	Environment of Formations
(1)	1	volcanic
(2)	3	swamp
(3)	5	deep marine
(4)	9	terrestrial

45 ______

Base your answers to questions 46 through 50 on the information below and on your knowledge of Earth and Space Sciences. Some questions may require the use of the ***2024 Edition Reference Tables for Earth and Space Sciences****.*

Atmospheric Carbon Dioxide

The chemistry of the oceans and coastal regions around the world are particularly vulnerable to changes in greenhouse gases. As atmospheric carbon dioxide emissions and polluted waters enter the marine environment, the waters become more acidic and less suitable for certain species. The full extent of this threat to marine ecosystems by human activity is not well understood, but proactive efforts are needed to protect the well-being of New York State's marine ecosystems. The graph below shows changes in atmospheric CO_2 levels in parts per million (ppm).

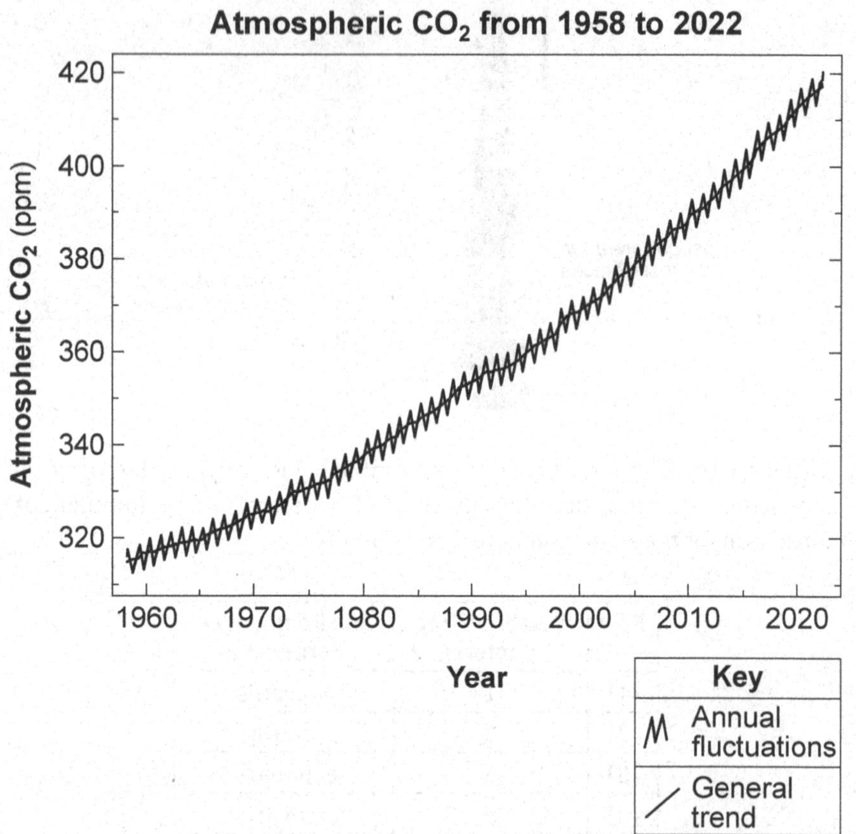

The graph below shows the change in global air surface temperature compared to the long-term average (0.0°C temperature anomaly). The line represents the general trend in the data points for the indicated time period.

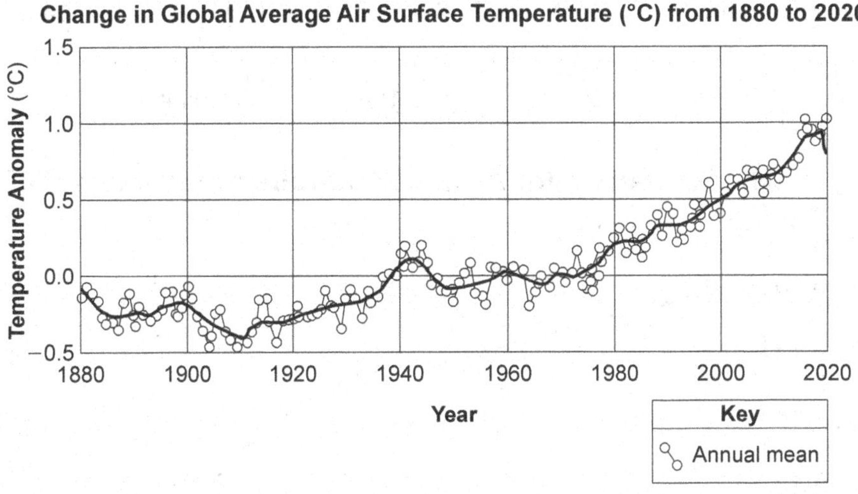

46. Describe how atmospheric CO_2 levels have affected the flow of energy out of the atmosphere *and* describe how this change in energy flow has affected the global climate. Use evidence from *both* graphs in your description. [1]

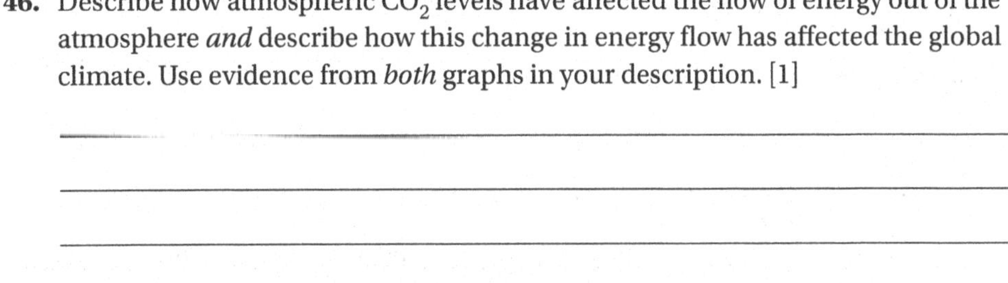

The model below shows the pH scale (0–14) that is a measure of the acidity or alkalinity of water. The graph below shows some information about the pH of ocean water.

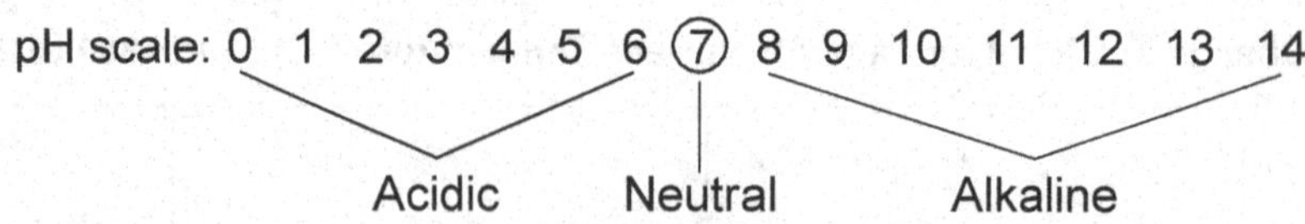

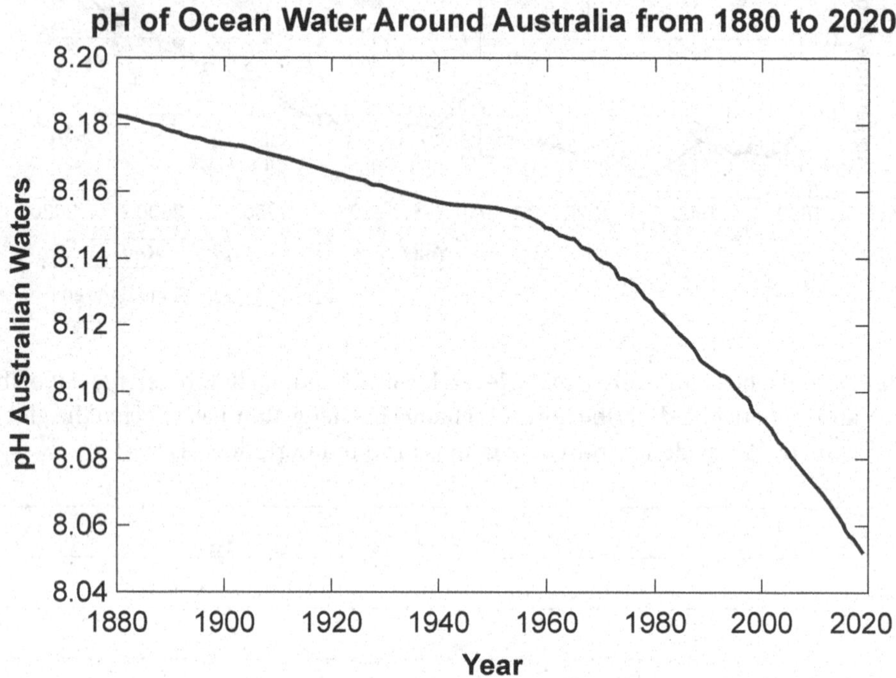

47. The change in the atmospheric carbon dioxide level from 1958 to 2020 has caused the waters around Australia to show

(1) a decrease in oceanic pH and an increased acidity
(2) a decrease in oceanic pH and a decreased acidity
(3) an increase in oceanic pH and an increased acidity
(4) an increase in oceanic pH and a decreased acidity

47 ______

The acidity of ocean water is determined by the relative amounts of H^+ ions. The models below show the ocean acidification process for two different time periods. The thickness of the arrows indicates relative amounts of substances in the atmosphere and oceans.

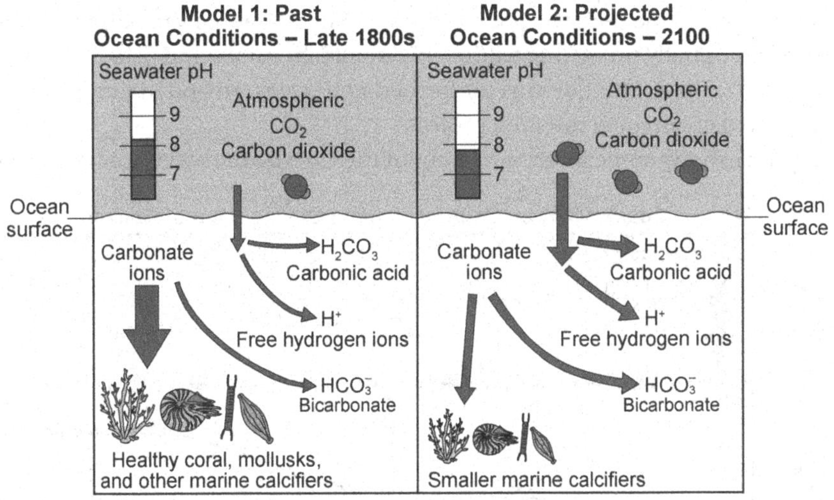

48. From the time in *Model 1* to the time in *Model 2,* determine if the atmospheric CO_2 entering the ocean and the ocean carbonate ions are projected to increase, decrease, or remain the same. [1]

Atmospheric CO_2 entering ocean: ______________________________

Ocean carbonate ions: ______________________________

49. Use the information in the models to explain how the projected (2100) availability of carbonate ions will change *and* will affect the marine calcifiers. [1]

____

____

____

Mineral weathering, which takes place over geologic time scales, is one of the main mechanisms Earth uses to recycle carbon dioxide. In a process called Coastal Carbon Capture, a common volcanic mineral, olivine, is used to limit the impacts of CO_2 on the hydrosphere. In this process, olivine is mined from Earth, ground into sand-sized particles, and then transported to coastal regions where it is spread along shorelines. Once in the water, the olivine sand will absorb hydrogen ions, gradually making the water less acidic and potentially protecting wildlife like shellfish, corals, and fish.

In July 2022, this technique was performed at the eastern end of Long Island, New York, as part of an ongoing research project.

The model shows some information about the Coastal Carbon Capture process.

Coastal Carbon Capture: Adding Olivine Sand to Oceans

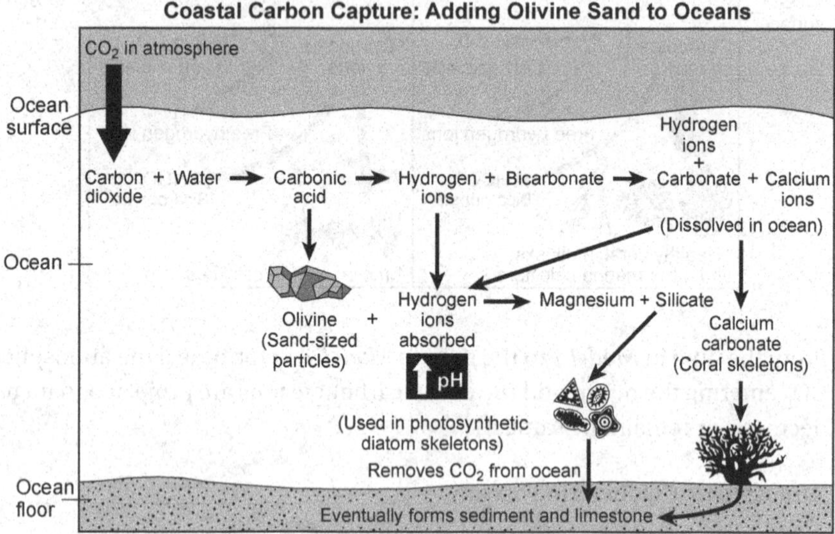

50. Which claim correctly summarizes how the Coastal Carbon Capture process is a solution that will reduce the impact of ocean acidification?

(1) Olivine will lower the pH of the ocean and decrease CO_2 levels.

(2) Olivine will increase the pH of the ocean and decrease ocean CO_2 levels.

(3) Olivine will lower the pH of the ocean and increase ocean CO_2 levels.

(4) Olivine will increase the pH of the ocean and increase ocean CO_2 levels.

50 ______

Answer Explanations August 2025

1. **Sample Response:** Complete the nuclear fusion model.

$$\text{Deuterium (A)} + \text{Helium-3 (D)} \xrightarrow[\text{Fusion}]{\text{Nuclear}} \text{Helium-4 (C)} + \text{Energy (B)} + \text{Neutron}$$

Explanation

In the Sun's core, lighter nuclei fuse to form helium-4, releasing a free neutron and large amounts of energy.

2. **(1)** Energy is generated by nuclear fusion in the core. It then passes outward through the radiative zone (by radiation) and the convective zone (by convection) before being emitted at the photosphere as electromagnetic radiation.

Wrong Choices Explained

(2) The photosphere is an emitting layer, not a site of energy production; the core produces energy.

(3) The radiative zone only transports energy outward; fusion does not occur there.

(4) Energy cannot bypass the interior; it must traverse the radiative and convective zones before escaping into space.

3. **Sample Response:** Total solar irradiance rises during sunspot maxima and falls during sunspot minima. Peaks in irradiance align with peaks in sunspot number; troughs align with minima.

Explanation

Sunspot maxima coincide with brighter active regions and higher net output, raising disk-integrated irradiance.

4. **(4)** Solar sunspot numbers follow an approximately 11-year cycle. The correct graph projects a continued sequence of maxima and minima through 2040 and is consistent with the historic pattern.

Wrong Choices Explained

(1) This graph does not reflect the repeating 11-year cycle evident in observations.

(2) This graph places peaks and troughs inconsistently with the expected cadence of the solar cycle.

(3) This graph shows the wrong frequency and amplitude relative to the known cycle.

5. **(3)** Late in stellar evolution, core temperatures rise enormously, enabling fusion of progressively heavier elements up to iron; elements heavier than iron require a supernova. Thus, heavy-element production is associated with late stages at very high temperatures.

Wrong Choices Explained

(1) Lighter elements (H→He) form in earlier stages, not later ones, in a star's life.

(2) Supernovas create the heaviest elements, not lighter ones.

(4) Heavier elements require higher, not lower, temperatures.

6. **Sample Response:**

Property of water: Water expands by approximately 10% upon freezing, exerting pressure in pores and cracks of concrete.

Relative amount of deformation: Samples with rebar deform less than samples without rebar; steel reinforcement reduces cracking under freeze-thaw cycles.

Explanation

Freeze-thaw cycles wedge open microcracks as water turns to ice; rebar constrains deformation and distributes stress.

7. **(3)** Surface water dissolves minerals from concrete and transports the dissolved load a short distance. Then those minerals precipitate to form calthemite straws beneath the structure.

Wrong Choices Explained

(1) Evaporation removes water but does not dissolve minerals; condensing minerals is not a geologic process.

(2) Minerals do not melt or freeze in this context; deposition arises from precipitation out of solution.

(4) Absorption and expansion do not produce tubelike secondary precipitates.

8. **(2)** Hydroelectric generation uses a renewable resource (flowing water) and avoids burning fossil fuels, thereby reducing atmospheric CO_2 emissions and helping to mitigate global climate change.

Wrong Choices Explained

(1) Hydropower does not increase atmospheric O_2 in a way that drives climate change.

(3) Hydropower does not meaningfully increase atmospheric water vapor/humidity.

(4) Hydropower does not decrease water vapor or create more arid local climates as stated.

9. **Sample Response:** Diverting Niagara River water through tunnels to power plants reduced discharge over Horseshoe Falls. With less flow crossing the lip, erosive power declined and the rate of retreat decreased.

Explanation

Erosion rate varies directly with water volume and velocity over the brink; diversion lessens both.

10. **(1)** Hydroelectric shows the greatest energy return on investment (EROI) among the listed sources. Its EROI is roughly five times higher than that of the best nonrenewable energy source, indicating hydroelectric's comparatively easy, cost-effective energy production.

Wrong Choices Explained

(2) The largest EROI of a nonrenewable energy source is not equal to the EROI of the smallest renewable energy source. The EROI of hydroelectric clearly exceeds that of nonrenewable energy sources.

(3) Coal's EROI is not the same as wind's EROI.

(4) Renewable energy sources do not uniformly have half the EROI of nonrenewable energy sources. The EROI of hydroelectric exceeds that of nonrenewable energy sources.

11. **(2)** If the universe began in a hot, dense state, it would have released energy uniformly that would have cooled and stretched to microwave wavelengths—cosmic microwave background radiation (CMB)—that would be detectable in all directions. That is exactly what is observed.

Wrong Choices Explained

(1) The CMB found in all directions does not imply multiple explosions; it supports a single hot Big Bang.

(3) CMB is not produced by every star or galaxy; it is a relic of the early universe.

(4) CMB is not energy from dying stars; supernovas do not explain uniform microwaves found in all directions.

12. **Sample Response:**

Difference in wavelength: Light from the first stars has been redshifted to longer wavelengths (visible and ultraviolet (UV) light has stretched toward the infrared (IR) range).

How the universe changed: The redshift indicates space itself has expanded since the light was emitted, increasing wavelengths.

Explanation

Cosmological expansion lengthens photon wavelengths:

$$\lambda_{\text{observed}} = (1 + z)\lambda_{\text{emitted}}$$

13. **(4)** Methuselah contains trace metals (elements heavier than helium). The very first stars lacked such metals; these stars were approximately 75% hydrogen and approximately 25% helium. Thus, Methuselah cannot be among the first stars.

Wrong Choices Explained

(1) The first stars were not 25% iron; they were nearly metal-free.

(2) The first stars had fewer heavier elements than Methuselah, not more.

(3) Methuselah is not simply 50% hydrogen and 50% helium; it contains trace heavier elements.

14. Sample Response:

A: less

B: greater

C: less

Explanation

The Hubble Space Telescope's much smaller orbital radius means a stronger gravitational field at its location; stronger gravity implies a shorter orbital period than the Moon.

15. (1) Closer satellites orbit faster. The Hubble Space Telescope (HST), at an altitude of about 540 km, orbits much faster than the Moon, which has an altitude of about 384,400 km. The correct graph shows HST completing two orbits in less time than the Moon.

Wrong Choices Explained

(2) This graph incorrectly shows the Moon faster than HST.

(3) This graph reverses the expected speed-distance relationship.

(4) This graph misrepresents the number/timing of orbits for HST versus the Moon.

16. (4) At approximately 30°N and 30°S ("horse latitudes"), air is typically descending (subsiding) within the Hadley cell and then diverging at the surface, yielding light winds, clear skies, and little precipitation.

Wrong Choices Explained

(1) Air is not ascending here; it descends from above.

(2) Ascending motion is characteristic of the Intertropical Convergence Zone (ITCZ), not the "horse latitudes."

(3) Convergence at the surface accompanies rising motion; here the air diverges.

17. **(4)** From the northern "horse latitudes," the prevailing westerlies steer air masses toward the northeast (from southwest to northeast), consistent with cT movement toward approximately 60°N.

Wrong Choices Explained

(1) Northeast trade winds blow from northeast to southwest in the tropics, not in the mid-latitudes.

(2) Prevailing westerlies do not move cT air masses from northeast to southwest.

(3) Northeast trade winds do not transport cT air masses from southwest to northeast at those latitudes.

18. **Sample Response:**

Code for Air Mass *A*: cP (or cA)

Code for Air Mass *B*: mT

Weather conditions for the next few hours at the residence:

Weather Conditions	Increases	Decreases
cloud cover	X	
chance of precipitation	X	
air temperature	X	

Explanation

A warm, moist mT air mass advancing toward a cold, dry cP air mass raises both clouds and the precipitation risk along the boundary. Temperatures also trend upward as warmer air approaches.

19. **(1)** A stronger Bermuda high drives steering flow farther south and then west, tending to keep a hurricane on a more southerly track before turning west. Hurricane 2 shows this south-then-west path under a strong, clockwise high.

Wrong Choices Explained

(2) The track of east, then north contradicts the shown westward steering around a strong high.

(3) Winds around highs move clockwise in the northern hemisphere, not counterclockwise.

(4) This choice combines the wrong steering with an incorrect circulation sense.

20. Sample Response: At 12 hours after landfall, the maximum winds are stronger than at 24 hours. Because falling wind speeds correspond to weakening pressure gradients, the atmospheric pressure must be higher at 24 hours than at 12 hours.

Explanation

As tropical systems weaken over land, pressure rises while sustained winds decrease.

21. (3) Many asteroid water samples have deuterium-to-hydrogen ratios (D/H) close to Earth's oceans, supporting asteroids as a plausible major source of Earth's water.

Wrong Choices Explained

(1) The giant planets are not direct sources for the water in Earth's oceans. These planets have D/H ratios less than that of Earth's oceans.

(2) Comets show a range of D/H ratios. The D/H ratios of comets alone do not best match that of Earth's oceans.

(4) Enceladus's water D/H ratio is not the primary match for the D/H ratios of Earth's oceans.

22. (2) Volcanic outgassing released water vapor that condensed and accumulated to form Earth's early oceans; the model shows water reaching the oceans from interior outgassing.

Wrong Choices Explained

(1) Not all early volcanic gases escaped to space; many were retained.

(3) Early volcanic outgassing was dominated by water, carbon dioxide, and nitrogen. Hydrogen and helium were largely lost to space.

(4) Volcanoes do not directly emit oxygen used by plants; oxygen accumulation later came primarily from photosynthesis.

23. Sample Response: Stromatolites carried out photosynthesis in early oceans, removing carbon dioxide from the ocean-atmosphere system and releasing oxygen, progressively increasing atmospheric oxygen over billions of years.

Explanation

The model shows oxygen rising while carbon dioxide decreases as photosynthetic life proliferated.

24. Sample Response:

Claim #1: As the percent of plant cover increases, runoff will decrease (and as the percent of plant cover decreases, runoff will increase).

Claim #2: As the percent of plant cover increases, the amount of soil eroded during a rain event will decrease (and as the percent of plant cover decreases, the amount of soil eroded during a rain event will increase).

Explanation

Vegetation intercepts rainfall, enhances infiltration via roots, and stabilizes soil, reducing overland flow and sediment yield.

25. (2) Well *B* taps more porous, more permeable sediments than does Well *C*, allowing a faster water supply rate at Well *B*.

Wrong Choices Explained

(1) Lower porosity with higher permeability is internally inconsistent and does not match the cross section.

(3) Higher porosity with lower permeability at Well *C* would still yield slower well response than at Well *B*.

(4) Lower porosity with higher permeability at Well *C* contradicts the observed stratigraphy and expected rates.

26. Sample Response:

Water table depth: A prolonged drought lowers the water table. (The water table falls farther below the surface.)

Effect on Well *A* and Well *B* residents: Well *A* users (shallow Upper Glacial aquifer near the coast) are more likely to lose supply or experience salt water intrusion than Well *B* users (deeper Magothy). So users of Well *A* are affected more severely.

Explanation

Reduced recharge drops wellheads; shallow coastal wells are first impacted by drawdown and salt water encroachment.

27. **Sample Response:**

Letter of model: *C*

Geologic process: Convergence and subduction recycle oceanic lithosphere; partial melting above the subducting slab produces magma that rises—cycling matter from the crust/mantle back toward the surface.

Explanation

Model *C* shows a subduction setting that generates magma—an explicit mechanism for matter cycling.

28. **3** The uplift of granite and lower marble is constructive because it builds topography and exposes new crustal sections. Classifying it as constructive correctly complements the other entries in the student's chart.

Wrong Choices Explained

(1) Shrinking of the Trans-Adirondack Basin is destructive, because it reduces land area and involves crustal collapse, not construction.

(2) Initial faulting along the basin margins is destructive. Faulting breaks and displaces rock, weakening the crust. It represents crustal breakdown rather than building new landforms, so it cannot be considered constructive.

(4) Filling with upper marble (a product of metamorphism and deposition) is constructive, because it adds new material and builds up layers, but this is not the correct stage referenced in the question.

29. **(2)** After initial basin formation and volcanism, continued seafloor spreading widens the ocean basin before later compression/metamorphism—so the missing second stage is continued formation and widening of the ocean floor.

Wrong Choices Explained

(1) Quartzite does not form directly from lower marble.

(3) Fault motion beneath the basin is not the defining second stage.

(4) Divergence widens (not narrows) the basin during this interval.

30. **(1)** Sulfide-based deposits associated with zinc and copper form in back-arc basins and at mid-ocean ridges (volcanogenic massive sulfides), the two tectonic settings indicated in the model.

Wrong Choices Explained

(2) Fore-arc basins and magmatic arcs relate more to other deposit types (e.g., porphyry copper in arcs).

(3) Granitic plutons host porphyry systems, not typical volcanogenic massive sulfide (VMS) zinc-copper deposits.

(4) A generic subduction zone without the back-arc spreading center misses volcanogenic massive sulfide (VMS) settings.

31. **(3)** High tides about 12 hours apart reflect Earth's rotation carrying a location through two tidal bulges each day (one facing the Moon and one on the opposite side). The cyclic pattern is evidence of rotation-driven tidal forcing.

Wrong Choices Explained

(1) The Moon's rotation does not cause the twice-daily cycle at a point on Earth.

(2) Earth's revolution around the Sun produces annual changes, not tidal bulges twice a day.

(4) The Moon's revolution affects timing, but the twice-daily highs are due to Earth's rotation through the bulges.

32. **Sample Response:**

Distance: 3800 cm

Explanation: The Moon recedes on average 3.8 cm/yr. Over 1000 years, that equals 3800 cm. This is tiny relative to the 384,400 km Earth-Moon distance, so the Moon's orbital period changes negligibly.

Explanation

A fractional change of approximately 10^{-8} in distance produces an imperceptible change in period.

33. Sample Response: As the Moon continues to recede, its apparent diameter will shrink. In about 700 million years, the Moon will be too small to fully cover the Sun's disk at perigee, making total solar eclipses impossible; only annular/partial eclipses will be able to occur.

Explanation

Apparent size is proportional to 1/distance. So the Sun's apparent size will still exceed the Moon's in that future epoch.

34. (2) The Moon's measured angular diameter changes more than the Sun's because the Moon's orbit is more elliptical than Earth's orbit around the Sun; therefore, the Moon's apparent size varies more over an orbit.

Wrong Choices Explained

(1) The change in the Moon's measured angular diameter is not smaller; it is greater.

(3) The Moon's orbit is not less elliptical than Earth's; it is more so.

(4) If the change in the Moon's measured angular diameter were smaller, it would imply a less elliptical orbit—contrary to observations.

35. Sample Response:

Sample: A

Explanation: Sample A retains approximately 49.31% of its original uranium-238, which is approximately one half-life. Because U-238's half-life is between 4.5 and 4.53 billion years, this supports a lunar age of about 4.53 billion years (Ga, giga-annum).

Explanation

Half of the parent radioactive element remaining implies that one half-life has elapsed.

36. **3** With the dam, seasonal discharge is smoothed and more controllable; the postconstruction hydrograph shows reduced extremes in river flow rates, enabling steadier and more manageable hydroelectric generation across the year.

Wrong Choices Explained

(1) River flow rates did not increase uniformly all year; variability decreased.

(2) Winter reductions in river flow rates do not maximize generation alone; overall control improved.

(4) River flow rates changed less after dam construction, not more, throughout the year.

37. **(4)** Reduced river flows lower the fresh water head and promote salt water intrusion into coastal aquifers, degrading habitats for many plants and animals and reducing biodiversity.

Wrong Choices Explained

(1) Lower water tables and reduced flows decrease (not increase) biodiversity.

(2) Lower flows typically lower, not raise, water tables in connected systems.

(3) Algal blooms from lower oxygen do not improve fish habitats; algal blooms degrade these habitats.

38. **Sample Response:** Before the dam, more sediment reached the coast and was deposited in deltas and nearshore zones. After construction, sediment is trapped in the reservoir behind the dam, reducing coastal sediment delivery and deposition.

Explanation

Reservoirs act as sediment sinks, starving downstream/coastal environments.

39. **(1)** Upstream of the dam, higher, steadier water levels expand calm spawning/rearing habitats, increasing opportunities for commercial fishing above the dam.

Wrong Choices Explained

(2) Dissolved oxygen often decreases downstream with reduced turbulence/flow.

(3) Upstream water levels do not decrease; reservoirs raise them.

(4) Downstream water quality generally declines with reduced oxygen; that hurts, not helps, fisheries.

40. Sample Response:

Amount saved: $31,076

Explanation

The damage amount avoided for a 24-inch flood is $87,326. Subtracting the elevation cost of $56,250 yields a net savings of $31,076.

41. (3) *Eurypterus remipes* belongs to the family Eurypteridae, is Silurian in age (about 420 mya), and used swimming appendages. The silhouette and bar on the cladogram place it in that interval and lifestyle.

Wrong Choices Explained

(1) Rhenopteridae, about 350 mya, and primarily walking do not match *E. remipes.*

(2) Pterygotidae, about 330 mya, and primarily walking are inconsistent with the state fossil.

(4) Mycteropidae and about 355 mya do not match the Silurian age or the family of *E. remipes.*

42. Sample Response: Deep-marine eurypterid families (e.g., Pterygotidae and Megalograptidae) show shorter temporal ranges than families in fresh water (e.g., Mycteropidae, Hibbertopteridae), which persist longer into the Devonian-Permian time periods.

Explanation

The cladogram bars indicate many lineages shifted from deep marine to shallow/fresh water before extinction, with clades in fresh water lasting longer.

43. (2) The extinction of many swimming eurypterid families occurs at the end of the Devonian period; after this event, most remaining eurypterids are found in fresh water environments.

Wrong Choices Explained

(1) The big extinction aligns with the end of the Devonian period, not the Silurian. Additionally, living in fresh water dominates afterward.

(3) Deep marine dominance after the event is not supported; living in fresh water becomes prevalent.

(4) The Silurian period is incorrect for this mass extinction signal.

44. **(2)** The paleomap shows New York, Morocco, and Venezuela located in a similar Devonian climate belt conducive to forest growth, explaining similar fossil forests on now-separate landmasses.

Wrong Choices Explained

(1) These three landmasses were not all directly adjacent; plate reconstructions show similar latitudes/climates instead.

(3) Characterizing them strictly as "northern hemisphere landmasses located above water" misses the climatic control.

(4) Not all sites were mountainous/tropical in the way stated; climate belt similarity is the key.

45. **(4)** The reconstruction indicates *Eospermatopteris* trees approaching 9 meters in height in a terrestrial (nonmarine) setting; the Gilboa fossils occur in fluvial/terrestrial deposits.

Wrong Choices Explained

(1) A height of 1 meter is far too short, and volcanic is not the depositional setting.

(2) A height of 3 meters and a swamp environment are inconsistent with the reconstruction and context.

(3) A height of 5 meters is inconsistent with the reconstruction, and deep marine settings do not preserve standing terrestrial trees.

46. **Sample Response:** As atmospheric CO_2 levels increased from approximately 320 ppm to over 410 ppm, more outgoing long-wave radiation was trapped, reducing energy flow out of the atmosphere and raising global mean surface temperatures by nearly 1°C over the period shown.

Explanation

Higher greenhouse gas concentrations strengthen the greenhouse effect and warm the climate system.

47. **(1)** Rising atmospheric carbon dioxide levels lower ocean pH, making the ocean more acidic. The graph shows a decrease over 1880–2020 in the pH of Australian waters, indicating increased acidity as atmospheric carbon dioxide levels rose.

Wrong Choices Explained

(2) Decreasing pH corresponds to increased acidity, not decreased.

(3) Oceanic pH did not increase; acidity increased.

(4) An increase in pH would imply less acidity, which is not observed.

48. **Sample Response:**

Atmospheric CO_2 entering ocean: increase

Ocean carbonate ions: decrease

Explanation

Greater CO_2 drives carbonic acid formation, shifting carbonate-bicarbonate equilibria and lowering carbonate ion availability.

49. **Sample Response:** With fewer carbonate ions, marine calcifiers (corals, mollusks, many plankton) struggle to build calcium carbonate skeletons/shells; growth and survival decline under projected 2100 conditions.

Explanation

Lower concentrations of the carbonate ion raise saturation thresholds, increasing the risk of dissolution and reducing calcification rates.

50. **(2)** Adding olivine promotes chemical weathering that consumes hydrogen ions, raising pH (makes the ocean less acidic). It also draws down dissolved CO_2 (as bicarbonate/carbonate), thereby reducing ocean acidification impacts.

Wrong Choices Explained

(1) Olivine increases, not lowers, ocean pH (makes the ocean less acidic). Olivine also reduces ocean CO_2 levels.

(3) Olivine does not lower pH or increase ocean CO_2 levels.

(4) Olivine does not increase either pH or ocean CO_2 levels.